THE BIRDS IN THE OAKS

H
HEYDAY
1974
2024
50

THE BIRDS IN THE OAKS

SECRET VOICES OF THE WESTERN WOODS

JACK GEDNEY

ILLUSTRATIONS BY ANGELINA GEDNEY

BERKELEY, CALIFORNIA

Library of Congress Cataloging-in-Publication Data

Names: Gedney, Jack, 1989- author. | Gedney, Angelina, illustrator.
Title: The birds in the oaks : secret voices of the western woods / Jack Gedney ; illustrations by Angelina Gedney.
Description: Berkeley, California : Heyday, [2024] | Includes bibliographical references.
Identifiers: LCCN 2024008533 (print) | LCCN 2024008534 (ebook) | ISBN 9781597146593 (hardcover) | ISBN 9781597146609 (epub)
Subjects: LCSH: Birds--California--Popular works. | Oak--California--Popular works. | Birds--Ecology--California--Popular works. | Oak--Ecology--California--Popular works. | Birds--Habitat--California--Popular works.
Classification: LCC QL684.C2 G429 2024 (print) | LCC QL684.C2 (ebook) | DDC 598.09794--dc23/eng/20240405
LC record available at https://lccn.loc.gov/2024008533
LC ebook record available at https://lccn.loc.gov/2024008534

Cover Art: Angelina Gedney
Cover Design: Marlon Rigel
Interior Design/Typesetting: Marlon Rigel

Published by Heyday
P.O. Box 9145, Berkeley, California 94709
(510) 549-3564
heydaybooks.com

Printed in East Peoria, Illinois, by Versa Press, Inc.

10 9 8 7 6 5 4 3 2 1

For my parents

CONTENTS

Preface ix

THE HEART OF THE OAK WOODLAND

1. The Soul of the Oaks: *Oak Titmouse* 1
2. The Prober of the Furrows: *White-breasted Nuthatch* 13
3. The Kinglet and the Greenlet: *Ruby-crowned Kinglet and Hutton's Vireo* 25
4. The Relatability of the Absurd: *Acorn Woodpecker* 39

THE BIRDS BENEATH THE OAKS

5. Discontented Shadows: *Spotted Towhee* 51
6. The Gentleness That Binds: *California Quail* 63
7. A Bird at Our Level: *Bewick's Wren* 75
8. The Little Tailor: *Bushtit* 87

VISITORS FROM THE PINES AND FIRS

9. A Fire Rises: *Northern Flicker* 99
10. The Antidote of Fear: *Chestnut-backed Chickadee* 111
11. The Winter Thrushes: *Hermit and Varied Thrush* 123
12. A Light in Winter: *Townsend's Warbler* 137

BETWEEN OAK AND SKY

13. Grayness Overcome: *Western Bluebird* 149
14. The Living Wind: *Band-tailed Pigeon* 161
15. The Falcon of the Oaks: *American Kestrel* 173
16. The Sky Elf: *Violet-green Swallow* 185

Acknowledgments 197
Notes 199
About the Author 217
A Note on Type 219

PREFACE

This is a book about the birds of California's oaks. I have always loved these trees, and the cast of characters who make their homes among their twisting limbs. I have walked and camped, studied and worked, lived and written beneath their branches, always accompanied by these voices in the trees.

I have trudged up and down the hills in search of acorn woodpeckers laughing in the scattered groves of oaks, bluebirds hovering over the flower-covered slopes, and winter warblers glowing in the dark woods along the creeks. I have eased into wakefulness as quail announced the morning and heard the rush of wild pigeons as they stormed by overhead. I have welcomed spring as one after another the birds declared the nesting season: titmice, towhees, and vireos successively joining the chorus of the woods. I have awakened before dawn to the sound of swallows singing in the starlight.

"Omit the negative propositions. Nerve us with incessant affirmatives. Don't waste yourself in rejection, nor bark against the bad, but chant the beauty of the good," wrote Emerson. "Set down nothing that will not help somebody." I have found few activities more affirming than walking among the oaks with my ears open to the birds. In these pages, I hope to convey some fragment of their chanting, that it may be of help to you.

CHAPTER 1
THE SOUL OF THE OAKS

Oak Titmouse, *Baeolophus inornatus*

> Titmice always see to it you are not lonely as you walk through the woods.
>
> —Neltje Blanchan, *Bird Neighbors*, 1897

What is the antidote to loneliness? Conversation.

Birds alleviate emptiness, and fill the air with sound. This is true of all birds to some degree. Few places are entirely without the speech of beaks and feathers. But it is not mere sound that transforms solitude into companionship. We can hear the birds' most public pronouncements—the songs of spring or the chatter of flocks—and fail to find our isolation mended. But there are a few birds who make me constantly aware of their individual existences, who paint themselves with every word as beings connected and at home. Nothing joins what has been severed like the speech of one to one.

Oak titmice are the birds among the oaks who never cease from speaking, whose constant back-and-forth continues in every month of the year. This is because the titmice of the oaks are the most monogamous of their kind, partnering in their first autumn and not parting until their last. Their conversation does not end, because there is always someone to listen and reply. Their songs are the first to start, because together they have fixed their borders long before the spring arrives. In the sandpaper of their calls there is a friction you can hold onto. And in the clear ringing of their songs I hear spring burst forth like flowers from winter's silent soil.

The Titmouse Song

Spring begins when the titmice sing.

A few outliers begin their nesting music even earlier—hummingbirds and great-horned owls are heard regularly in December, when titmice are only tentatively experimenting with song. But by the end of January, I hear these gray and crested birds singing every morning, inaugurating spring in coastal California, no matter what the calendar may say. The woods are leafing out and newly strewn with flowers, but most of the dawn chorus is still uncertain and unconvincing.

The titmouse song convinces, declares the dawning of the nesting season without doubt or hesitation. I ride out on my oak-lined street on a still-wintry morning and hear that new-year reveille. I step out at the day's beginning when even California's grasses crunch with frost and feel the warmth of this music pour into me like strong, dark coffee, insistently repeating: *life is no longer sleeping.*

This song stands out, not for subtlety or sophistication, but for clearness and forthrightness. The classical form involves three-to-eight repetitions of a two-note couplet: a higher and a lower pitch rapidly alternate in a pattern sometimes rendered *peto peto peto*. (Titmice, especially the tufted titmouse of the East, have been known as "peto," "Peter," or "Pete bird" for this song.) Sometimes the two pitches are both distinct and fairly well balanced, but often the lower tone is reduced to a brief grace note or sliding entrance into the main pitch. And while couplets are typical, they are not the only possibility: oak titmice often repeat a single pitch in a slowish trill, or less frequently opt for three-pitched, triplet motifs.

In other words, titmice songs are not identical. Become familiar with the most familiar version, two-pitched and unhurried, and then embed the tone quality in your mind. Even in their least distinctive, single-pitched rendition, their voice rings in a uniquely rich middle register. Their song is not high, dry, and thin like the junco's gentle trilling; it is not preceded by a buzzy pickup like that of Bewick's wrens; and it is patient rather than propulsive like the rubber-band twanging of the spotted towhee.

Each male titmouse has a repertoire of ten or so unique songs. Titmice learn and repeat the songs of others around them, with neighboring males often counter-singing back and forth. Although we can't as a rule distinguish a bird's full repertoire without analysis of recordings, we can hear this: one titmouse sings, and another responds identically, giving his answer more force and focused challenge than would a more generic statement to the world. In one study on oak titmice, Keith Dixon fittingly rechristens these vocalizations with a more pungent epithet than our excessively harmonious word "song": "fighting notes."

I think of this vocal exchange as comparable to human face-offs in which two tough guys escalate tension with challenging restatements of each other's phrases—an *"Oh, yeah?" "Yeah!"* dynamic. Effective repartee relies on repetition. As my childhood hero, the Horace-quoting pirate doctor Captain Blood, once demonstrated, if a rapacious buccaneer shouts out, "You will not take her while I live!" the most piercing riposte is clearly, "Then I'll take her when you're dead!" I like to imagine counter-singing titmice vigorously shouting those words back and forth between the trees. It's good practice to avoid ambiguity when challenging someone to a duel.

In titmice quarrels, to be clear, females are not the immediate subject of dispute—the territorial boundary is—but they are the underlying evolutionary cause of conflict. The primary function of a nesting territory is to guard access to the nest and mate. As Dixon sums up in another study, "Song and fighting appear to be more intense when the female is present."

What was true for pirates is also true for titmice. The fundamentals of monogamy are much the same for the feathered and unfeathered, and there is little more fundamental to monogamy than jealousy. Birdsong is not strange to us, but sings meanings that we know.

Calls as Domestic Conversation

Titmouse song is clear and ringing, the defiant and defining song of early spring. But calls—the catch-all term for birds' simpler, year-round vocalizations other than songs—are another subject entirely. Titmouse calls are squeaky and scratchy, and their season has no end.

I hear them in the spring, calling between bouts of song to proclaim that they, more than the season's other singers, have already found the one they searched for. I hear them in the autumn, declaring that their bond is still enduring. And walking in the crisp, cool air when the black oak leaves have turned to gold upon the ground, I hear them, those scratchy voices never to be worn smooth, their friction sparking warmth when the earth lies cold and dormant.

Some calls serve as keep-in-touch notes, some as territorial statements, and some as alarm calls, with subtle degrees of meaning mostly beyond our grasp. But even when we cannot discern the finer details of *what* is being said, we can without too much difficulty perceive *who* is doing the saying: that raspy, scratchy tone of the titmouse voice is entirely distinctive. It is also distinctively endearing. "Raspy" and "scratchy" are not adjectives of ugly dissonance, but correlates of expressivity. Clear whistled voices lack texture. Give a kid a penny whistle and she may produce a nice, clear tone, but that's about all. Give the same child a kazoo, and she will make all kinds of funny sounds expressive of indignation, disappointment, depression, or cheerful curiosity. I may not always know what the titmice are saying, but they always sound like they're saying *something*. As they are.

The titmouse tone feels homey, relatable. It reminds me of the famously quirky voice of Jean Arthur, the great comedienne of 1930s Hollywood. Writing after her death in 1991, Dennis Drabelle anthologized her admirers:

> Pauline Kael summed up Arthur's voice as "wistful-husky . . . one of the best sounds in romantic comedies of the '30s and '40s." Another critic, Gerald Weales, wrote of the Arthurian "odd throaty quality that was a cross between sandpaper and a caress.". . . Andrew Bergman gave up on comparisons and simply called it "the most wonderful speaking voice anyone had ever heard."

To maximize your titmouse appreciation, I therefore recommend watching the classic comedies *Easy Living* and *The More the Merrier*, relishing the sound of Arthur's down-to-earth and husky sandpaper.

The qualities of the titmouse voice are not always the most celebrated of vocal traits, but I suspect that I am not alone in being more drawn to raspy squeakiness than to some clear, operatic soprano: such tones sound gentler, more approachable, and, for lack of a better word, more human. Not alien, mechanical, or divine, but familiar, embodied, grounded. People may notice the most extravagantly singing birds and admire their virtuosic musicality. But titmice don't just sing—they *talk*.

They issue constant warnings because they have someone to keep safe. They ward off intruders from their borders because their nation holds just two. And nothing makes them feel so strong as the constant reassurance that they do not fly alone.

A Nation of Two

One of my favorite pastimes is wandering through the woods alone, being as quiet and unobtrusive as possible, looking and listening without the ruckus and chatter that accompany a group. There's only one thing better: walking side by side with the one who knows me best, Angelina, my wife and the illustrator of this book. She knows the things I'm searching for and multiplies my vision and my hearing. She recognizes why I stop before I say a word.

That's how titmice live. Not alone and not in flocks, but in pairs. When I visit any stand of oaks—from the handful of trees scattered around the edges of my closest local park to the hills surrounding town, blanketed in miles of forest—I seldom fail to find their familiar conversation. Today I am in the heart of titmouse country, a rich mixed woodland where live oaks, black oaks, and valley oaks intermingle with buckeyes and madrones. Daylight shines through, softened but not shut out, in an inviting middle point between the darkness of the forest and the unrelenting sunshine of the more exposed and open hillsides.

An assertive gray form moves through the canopy with confident deliberation, free from the nervous restlessness of kinglets or bushtits, who glean only from the surface and find little reward in slowing down. The scratchy-voiced bundle of crest and feathers passes from branch to branch, favoring the middle ground between the fine twigs favored by the foliage-gleaners and the furrowed trunks plumbed by long-billed creepers and nuthatches. And as she stops to pry more deeply or to chip away at a piece of bark, I search among the leaves and find her companion close at hand.

The contents of birds' thoughts and feelings are often opaque to us unobservant giants. A little attention starts to reveal *events*—particular moments when fear and alarm, aggression or conflict, spark a relatively dramatic reaction of movement or vocalization. But once you begin to notice when things *happen*, slow down and look more closely—now you are ready to perceive what is always *going on*. Watch and listen to the titmice when nothing else intrudes and you will see their life of constant sharing, the side-by-side existence of two who live together.

It is this quiet baseline of coexistence that is the strongest indicator of what I have previously called a "nation of two" (borrowing a phrase from Kurt Vonnegut's *Mother Night*). Sometimes you will perceive little overt expression of this bond beyond the presence of precisely two birds in close proximity, foraging together. You might see more conspicuous interactions of pairing during the nesting season: mutual preening, mate feeding, copulation, shared visits to a nest site. But what defines a nation of two is not the peak of breeding fervor, shared by many birds, but what happens all year round.

For the majority of North American songbirds, pair bonds are temporary, dissolving with the conclusion of the nesting season as birds revert to isolation or gather into flocks. A minority maintain long-term monogamous bonds and defend a shared territory throughout the year, generally until the death of one of the two members. Oak titmice are among this general minority of birds, and are even more exceptional among their closest relatives: the other fifty-eight or so members of the Paridae family all appear to flock in winter, to greater or lesser degrees, at least among the well-studied titmice and chickadee species of Europe and North America. Oak titmice never gather into larger flocks, because two birds are enough: this bird is *the* monogamous parid.

Presumably thinking of petite adorability, Dickens addressed his love letters to his "dearest titmouse," among his many epithets of enthusiasm, before signing off with a quantity of kisses somewhere between "several zillions" and "an unlimited number." I support the revival of "titmouse" as a term of endearment commensurate with this volume of kisses, and as a term especially appropriate in California, home of the world's most monogamous titmouse.

Multiple factors likely contribute to the singularly enduring attachment of oak titmice. The first great enabler of year-round monogamy is year-round residency—California's temperate climate with plentiful food even in winter makes migration relatively less common. "Of all tits so far studied Plain Tits are the most sedentary," says John Gibb of Oxford, employing the old common name in its Anglicized form. Our titmice are the most sedentary because they live in the nicest place.

A second factor is the nature of titmouse food sources: during winter, a significant portion of their diet comes from acorns and other shelf-stable, cacheable seeds—storing food for later is more valuable when there are no flock-mates to steal it. A third benefit to year-round pairing comes from the competition over tree cavities used for nesting. Titmice nest in natural or woodpecker-excavated holes, which are often a limiting factor in the widely spaced deciduous oak woodland that makes up their core habitat. A key titmouse strategy is therefore to claim a nest site early and defend it vigorously; already-paired resident birds have a significant advantage over those that need to spend the early phase of the nesting season establishing new pairs or territories.

I don't know of any easy way to assess the relative importance of these three strategic correlations to the titmouse preference for year-round monogamy. But all three of these alignments have been noted among birds in general and all apply to oak titmice to a greater degree than to the median temperate songbird: permanent pairs are more common among non-migratory birds, food-caching birds, and cavity-nesters than among migratory birds, non-caching birds, and cup-nesting species that face less fierce competition for nest sites.

Amenable year-round climate and foraging opportunities, prolific production of readily cacheable food, and present but not overabundant nesting cavities—these are the traits of the California oaks. Other parids flock in winter, but not the one that settled among these trees. Even the nearly identical Juniper titmouse of the eastern Sierra sometimes flocks in winter, as does the Woodhouse's scrub-jay who shares their home among the pines and junipers. But the scrub-jays of the oaks, like the titmice of the oaks, are nations of two.

California's oaks breed possessiveness of place. To be the most sedentary does not imply a lack of boldness or of vigor. To be the most sedentary is to have the strongest roots.

The Oaks' Soul

This crested parid is the voice and soul of the oaks.

—Dave Shuford, *The Marin County Breeding Bird Atlas*

I leave the trail's winter mud and bend the new-grown grass to walk beneath the trees, widely spaced across the gentle slope like fantastically irregular umbrellas. Woodpeckers fly between the oaks, which rise like islands from the sea of waving green to offer acorns and insects, leaves and flowers, galls and mistletoe, nest sites and granaries. We live on the ground and see the trees as disruptions of our default horizontal. But for many birds the *trees* are the world they live in, a world of three dimensions in which the ground is the least interesting part.

I lie down beneath the trees to watch the spreading branches overhead. These valley oaks are thick limbed like Polykleitan heroes, massive arms protruding from massive trunks, each line thickened and grown strong from centuries of reaching. Looking upward from the earth, disassociated from my customary standing orientation, I realize the relative unimportance of my old sense of weight and gravity as I watch the titmice make their rounds. To live among the oaks is to live among a million surfaces, a world of endless angles so much richer than our flatness, a world of changing leaves and catkins rich with pollen that emerge to festoon the birds' home with great swathes of color. Among all these wonders, small forms dart like children in the most extravagantly designed of playrooms.

The soul of the oaks pauses, crest relaxed, on one of those deep-furrowed limbs, an edifice as solid to her as any building we construct. She explores within the cracks and crevices, hammering away at a concealing ridge of bark to extract some unidentifiable arthropodal bundle. Her partner is close behind, tending to the branch tip buds as he seeks the emerging flowers or the insects that are drawn to them. The first bird flies to the ground and then back to the branches, where she uses her feet to hold whatever it is she's found below—some seed or brittle-shelled invertebrate—as she hammers with her beak.

Oak titmice live among the oaks, on their branches, twigs, and trunks. They find food fallen in their shade and nest sites hollowed in their wood. The prime habitat of the gray-crests are California's deciduous oak woodlands, relatively dry and open communities dominated by trees like valley, blue, and black oak, where they search for food in both the foliage and on the thickly furrowed bark. These furrows represent a significant foraging opportunity for titmice, who chip away at the bark-walled crevices with their sturdy beaks. This furrowed quality of the deciduous oaks stands in contrast to the smoother-barked live oaks, which hold fewer bugs in woody crannies and therefore present more opportunities to the lightweight foliage specialists like chickadees and vireos than to the comparatively heavy titmice. Titmice still live among the live oaks, but at lower densities than they are found in woodlands with a mix of deciduous species.

That sturdy beak is one of the oak titmouse's most anatomically distinguishing features among their cohabitants in the oak woodland. Some birds have tiny beaks (the foliage-gleaning chickadees, kinglets, bushtits, and warblers) and some have long, skinny beaks (the bark-probing creepers, wrens, and especially white-breasted nuthatches). Titmouse beaks are in-between: thick, solid, and useful for foraging in bark crevices without being overspecialized. The notable sturdiness of these tools indicates one of the most obvious points of titmice–oak alliance: they are the little birds most capable of eating acorns. These beaks are, to be sure, a flexible instrument, capable of cracking seeds and decapitating caterpillars that might intimidate the average bushtit. But while there are many small songbirds foraging for insects in the same tree as the titmice, none other than titmice routinely carry off intact acorns.

Not only do titmice consume acorns regularly, but they also cache them whole—a pastime usually reserved for much larger birds, notably woodpeckers and jays. Titmice are small enough to go almost everywhere and eat almost everything that the smallest gleaners do. But they are also just large enough to successfully intrude upon the fruitful niche of the acorn cachers, storing those rich, durable nuts in the soil like tiny jays. I delight in watching acorn woodpeckers as they defend their granaries, the clown-faced birds yammering at each other and at intruding crows and squirrels. But I get a special thrill of sympathy when a titmouse sneaks in to pluck an acorn from their stores, carrying off the bulky nut while her two-and-a-half-inch wings flutter furiously.

Here's what it comes down to: oak titmice love the oaks, and the oaks nourish them all year round, giving food and shelter. Acorns can be cached; insect eggs and larvae cache themselves in bark and gall to feed the gray-crests through the winter. Spring comes with buds, flowers, and fresh profusion of arthropods of all sizes, while a cavity carved in oakwood by an acorn-nourished woodpecker gives shelter to titmice mothers and their six white, hidden eggs.

Titmice are oaken birds, sturdy and enduring. Conifers have fine and filmy foliage, fit only for the lightest gleaners, while these birds cling to stronger stems and hack at deep-ridged bark. Other plants have brief and flashy fruitfulness, like summer's birds of passing splendor. Oaks give in every season, and the gray-crests never leave. All year round, you'll hear the conversation of those husky, rasping voices, the sound of the oaks' soul speaking in their rough and rustic speech.

Oaks are symbols of that which lasts, of slow-built strength and thick, furrowed bark that withstands both pest and fire. In the first autumn of her life, a titmouse finds the voice that matches hers, and together those two stand their ground like acorns taking root. They forage through the winter without a day of silence. And when spring comes and the black oaks loose their leaves of soft-cheeked scarlet, then those unfading voices ring out clearer and blow away the cold.

CHAPTER 2
THE PROBER OF THE FURROWS

White-breasted Nuthatch, *Sitta carolinensis*

Birds make many different sounds. Some are long, complex, and musical, but even simple calls can have a huge variety of tones—raspy like a titmouse, buzzing like a wren, mildly clucking like a quail, or groaning like a spotted towhee, just to name a few. In many ways, timbre or tone quality is more consistently useful than pattern when it comes to recognition: you are unlikely to mistake a violin for a tuba, even if each instrument plays only a single note.

Few voices are as instantly recognizable as that of the white-breasted nuthatch: they say *quank* in loud and carrying calls, the most nasal sound in the woods. Before the modern standardization of names, this bird was sometimes referred to as "the big quank," as distinguished from "the little quank," the red-breasted nuthatch, who does not live among the oaks. Alternative transliterations are plentiful: other authors describe this call as *yank, keer, kun,* or *art*, the last of which sounds perhaps mildly closest to my ear. Nuthatches in the East or interior West will often double up these notes, or sometimes run many together into a kind of trill, but in coastal California it is most common to hear individual *quanks* honking into space among the oaks. And so, when my wife and I hear the little high-pitched foghorns blow, we eschew the formality of "nuthatch." The name that comes to our lips is invariably "quank."

I hear them now, on this day on the cusp of spring, when green is just beginning to emerge from the treetops of the oaks. "*Art . . . art . . . art*," they announce, compact little packages of bird flying short-winged from one tree to the next. I duck under the canopy of the great, spreading tree, and look up, searching for the quanks. They are not in the outer foliage, still alive with the birds of winter, yellow-rumped warblers and ruby-crowned kinglets flitting among the thin branches. I trace the limbs with my glance, following the long ridges in the bark as they deepen on the thick, lichen-spattered trunk. At last I find a nuthatch, steadily creeping downward, blue-gray backed and capped in black, gripping the trunk with an ant's carelessness of gravity. When I *see* nuthatches, a different one of their names comes to mind: "bajapalos," pole-goer-downer, descender of the masts.

"Nuthatch" is a fine old word, but "quank" and "bajapalos" strike to the heart of what I encounter first: uniquely nasal callers and unmatched acrobats of descent. Like titmice, nuthatches live an arboreal existence, a constant immersion within the world of bark and branch, leaves and flowers. But the nuthatch world is more narrowly proscribed, their habits more specialized and consistent than those of titmice, who feed on weighty uncracked acorns, bark crevice hiders, underleaf lurkers, and even ground-based crawlers with nearly equal enthusiasm. Not all birds are such generalists, and a large part of the pleasure of learning the birds among the oaks is learning the unique niche that each bird fills, the inimitable connections between each bird and the trees.

Each bird finds her place, seeks and finds the gifts the oaks are offering *to her*. You have to follow closely to see what each receives. But the easy way to learn this lesson of ecological specialization is to start by studying an unconfusable bird, a bird who lives with unmistakable extravagance in its every motion, pose, and utterance. It requires subtle study to tease apart the niches of titmouse and chickadee, bushtit and warbler, kinglet and vireo. But it takes just a moment's observation to recognize the uniqueness of the intrepid goer-downer, the quanking nuthatch clinging to the deep and furrowed bark.

The Diversity of Oaks

Fully understanding the nuthatch and all the birds to come calls for some understanding of the oaks in which they live. These are not just any trees, but the dominant and crucial members of the woods of California.

Biologically, oaks are members of the genus *Quercus*, trees characterized by the large, capped nuts known as acorns, wind-pollinated flowers in hanging catkins, and a form that tends toward the wide and spreading, with thick trunks and limbs that twist and reach dramatically rather than growing in simple vertical columns. Ecologically, oaks stand out first of all on the basis of a simple claim to fame: they are the most abundant and successful trees in the world.

This preeminence is true in California as it is in many other places; until you reach some 8,000 feet of elevation in the mountains, nearly every woodland or forest community includes oaks in at least a supporting capacity, and often as the dominant trees. This omnipresence automatically confers importance upon oaks for all tree-dependent needs of birds, from nest cavities and safe roosting sites to flycatching perches and refuge from ground-based predators. They have a direct, genus-specific contribution to the lives of many birds in the form of their acorns, which make up central parts of the diets of jays, woodpeckers, and more. And their overall global dominance is probably a factor in what may be their biggest importance of all for many birds: they support more insects and arthropods than any other trees, with some eight hundred arthropod species so far documented feeding directly on California oaks.

Oaks are not a monolithic group, but one with important variations. California has eighteen different oak species, nine of which grow predominately as trees and nine of which are largely shrubs. To start clarifying your mental categorization of these trees, the most helpful initial distinction to make is between the evergreen or "live" oaks (such as coast live, interior live, and canyon live oaks) and the deciduous species (such as valley, blue, Oregon white, and black oaks). Each species has its preferred setting in terms of climate, elevation, topography, and so on, although there are many areas where multiple species will overlap, typically with other non-oak species as well.

Take the coast live oak, for instance. This is a very common tree along the coast, including where I live in the San Francisco Bay Area. Its evergreen leaves are comparatively small and stiff and bear sharply pointed bristles, while the bark of even mature trees retains a relatively smooth, gray surface, only deepening into shallow furrows with great age. Some birds specialize in associating with these trees, such as the Hutton's vireo, a year-round, insect-eating foliage gleaner, whose talent at searching for prey on the undersides of leaves would not serve as well in winter among deciduous trees. Other birds are most at home in cool and shady forests, breeding in largely coniferous forests to the north but then finding a winter refuge among the dense broadleaf woods of coast live oaks and bays—the hermit thrush, varied thrush, and Townsend's warbler fit into this category.

Then there are the valley oaks, to choose the mightiest representative of the deciduous oaks and one of the key nuthatch trees in California. These trees have deeply lobed leaves that are rounded and without bristly tips, exceptionally large acorns, and deeply furrowed bark, but it is their sheer size and grandeur that is most striking: valley oaks are one of the competitors for largest oak in North America, living for several hundred years and with maximum trunk diameters measured at over nine feet. On average, valley oaks are found in more open assemblages than coast live oaks, rarely forming the dense, continuous canopies regularly found among the evergreen trees. The most magnificent valley oaks tend to spread out in what is technically not known as woodland at all, but as oak savanna: a mostly open grassland habitat beloved of many of the aerial hunters discussed later in the book, including kestrels, bluebirds, and violet-green swallows.

If you want to know the oak woodland birds, it is invaluable to start knowing the oaks. Sometimes relatively subtle traits of trees, which most people never think of, prove to be decisive for the birds. Nuthatches prefer deciduous oaks: valley oaks in the well-watered lowlands, blue oaks on the dry hillsides, and Oregon white oaks to the north. What trait of these trees drives their preference? Deeply furrowed bark, bark with cracks and crevices in which food can hide—and be hidden.

The bark of live oaks is relatively smooth and offers few hiding places: while some insects and spiders may periodically traverse the trunk, there is no comparison with the bug-abundant bark of valley oaks and their other deciduous relatives. Valley oak bark has a million little niches, impenetrable to my clumsy fingers and even the short, stubby beaks of most birds. Into these crevices retreat innumerable mites, springtails, and barklice, pursued by bigger mites, spiders, and beetles. Caterpillars and other larvae emerge from eggs deposited between the protective ridges. This array of bark-dwelling arthropods forms a large portion of the white-breasted nuthatch's diet, especially in spring and summer.

But it's not just the food *found* within the furrowed bark that draws nuthatches to these trees. It's also the food they *put* there. Nuthatches, like several other year-round residents of the oaks, store food for use throughout the winter, primarily seeds and nut fragments. Acorn woodpeckers store their supplies out in the open, but lodge them tight within their nooks and defend them vigorously; scrub-jays bury acorns in the ground where only they can find them. Nuthatches have another method for protecting their stores, placing them deep within the ridges of the bark, tamping seeds down and out of reach of the shorter-beaked chickadees and titmice and sometimes concealing their treasures with further scraps of bark or lichen.

The oaks host many birds: birds that eat the oaks' buds and

acorns, birds that eat bugs that eat oaks, birds that eat bugs that parasitize oaks, birds that eat berries from plants that parasitize oaks—the list goes on. They are capacious trees, trees that define so many of California's landscapes both by their imposing physical forms and by the vast array of animals that find nourishment and shelter among their trunks and branches. And amid all this multitude of offerings, all these different ways that the contents of an acorn have to convert air and sunlight into wood and fruit and flower, the upside-down-bird finds her niche, clinging to the thick-ridged armor within which food is found in spring and saved to see the winter through.

And How the Nuthatches Use Them

Tonight, I'm camping in the rolling oak woodlands east of the bay, enjoying the pleasantly warm air of spring before the heat of summer turns the hillsides brown. I walked throughout the morning along creeks lined with valley oaks and sycamores, through passages of chaparral fragrant with chamise, and across flower-strewn grasslands dotted here and there with rock wrens and meadowlarks. But now I've reached the wooded hill where I will spend the night, where blue oaks with their shallowly lobed leaves and vertically fissured trunks surround and lean over my tent. The walking's done, camp is made, and my belly is full as the sun begins to set. But a nuthatch still *quanks* as she moves along an oak limb and continues down the trunk with hop after jerky hop.

Her bill is long, far longer than those of her peers among the woodland songbirds. I watch as she jabs that long lance between the ridges of the bark, presumably spearing some tiny creature lurking out of sight. A few moments later, she comes across some slightly larger but still unidentifiable prey, seemingly rather large for a single bite—this she stabs a few times, then wedges in an outer crevice to give it several more whacks with her bill, quickly dividing the creature into more appropriately bite-sized portions.

The deciduous oaks offer nuthatches acorns, an abundance of arthropod prey, and deeply furrowed bark that they are particularly well-suited to exploit for foraging, caching, and even as a sort of tool, wedging food into place among the crevices while they crack open shells or hammer apart larger items. Above, I looked at this trait in terms of the differences between trees: these thick-barked deciduous species are the ones that nuthatches choose. Now it is time to look more closely at the differences between birds: how exactly do the nuthatches access these rewards that others can't?

It's easy for a casual viewer to simply lump together all the small songbirds hopping about in the oak canopy and neglect to see their finer differences. Even within this subset of the avian world, however, you can start distinguishing a few generalized feeding guilds: the leaf gleaners (including chickadees, bushtits, kinglets, vireos, and warblers), the flycatchers (including bluebirds, phoebes, and assorted other flycatchers proper), the chiselers into soft and rotting wood (larger woodpeckers), and the bark gleaners (small woodpeckers, creepers, and nuthatches). In reality, many birds practice multiple foraging techniques, so these descriptions should be regarded as tendencies rather than absolute laws. But recognizing a bird's guild membership is the essential first step to appreciating its unique place: the white-breasted nuthatch is the Ridged-Bark Gleaner.

Nuthatches' first and most obvious qualification for this title is that they are more adept at travelling on the trunks of trees than any other birds—woodpeckers and creepers not excluded. The old bird writers, always ready to acknowledge a physical superiority, all point to this first of all among the nuthatch traits: "the acknowledged acrobat of the woods," "the most expert climbers," and "no other bird can compete with the nuthatches in running up and down a tree trunk" declare Messrs. Dawson, Pearson, and Forbush.

While the technique of woodpeckers and creepers depends on the use of the tail as an additional point of support, nuthatches instead use both feet, splayed apart like a knock-kneed Elvis, each one with a highly developed and well-clawed hallux, or rear toe. This has an advantage—woodpeckers and creepers are essentially always forced to move upward on tree trunks, while nuthatches can move in any direction, including straight down and headfirst. This probably helps them find prey that those traveling upward miss, and to store food in places that those rivals likewise overlook. So this is the nuthatch's simplest claim to fame: of all the birds in the oaks, this is the bird that has mastered movement on them.

Their next most unusual feature is their bills. They are long, swords rather than the stunted daggers most birds have. They are long, like the oversized chopsticks my Taiwanese mother cooks with, while other birds reach vainly into the wok with miniature Swiss Army knife tweezers. Chickadees, vireos, kinglets, and bushtits all have bills under a centimeter in length. Titmice, wrens, and Nuttall's woodpeckers are in the 1 to 1.5 centimeter range. But nuthatch bills regularly exceed 2 centimeters in length. This is the key feature that allows nuthatches to reach down into those furrows. Their bills are not just grasping chopsticks, however, but can also hammer with a woodpecker's directed force. If you have nuthatches coming to a feeder for whole sunflower seeds, you can watch them take one, carry it to a suitable holding location, and then split it open with a few efficient blows. Their bills are tongs, wedge, and hammer in one.

Being able to access the trunks of trees and reach the hidden cavities within the bark enable a further practice of enormous value: caching food. The oak woodlands of lowland California host an unusually large number of cachers, in part because the climate encourages year-round residency and in part because the oaks themselves provide such abundant cacheable food. For most birds, this means scatter hoarding, placing individual items of food in different secret locations. Nuthatches, jays, woodpeckers, titmice, and possibly chestnut-backed chickadees all cache food.

A basic ability to access the bark is necessary, but nuthatches' ability to approach from above rather than from below is better, allowing them to use nooks that the upward-travelling woodpeckers and creepers often miss. An ability to reach into those crevices with a long bill is necessary, but nuthatches' ability to hammer open shells and break nuts into smaller pieces is better, expanding the range of items that can be stored. The ability to remember storage locations is a minimal requirement, but nuthatches' trick of covering up insufficiently concealed food caches with bits of bark or lichen makes it more likely that those food items will still be there when they return. The oaks are a well-stocked pantry of acorns, flowers, and insects. And the trees are not just the pantry *contents*, but also the shelves and bins and baskets that store the food and keep secure all the nuthatch's prudent savings.

There is a host of birds that live upon the oaks, finding food within their branches and shelter in their solid trunks. The magnanimous oaks welcome their dependence, forming lines of connection we can witness and appreciate. In our human world of purchasable conveniences of invisible origin, it's easy to slip into thinking that we operate apart, that we have no vital need for the rest of nature. It is much more difficult to make that error with birds, especially this troop of birds that together live upon the oaks, these trees like friendly giants bearing kinglets on one arm and vireos on another, with creepers, woodpeckers, and nuthatches clinging to their chest and back.

The story of the oak-borne birds is this: trees convert the invisible sustenance of the air and soil into great fortresses and gardens, enormous structures of branch and bark, leaves and flowers that offer food and shelter to thousands of creatures, each one with its proper place. This is the pattern of the world, of course, of the spotted towhees extravagantly kicking in the litter, of the flickers sending their improbable tongues in search of ants. The tale of evolution is this unendingly rich filling of every niche with astounding diversity.

When our field of view is narrower and we look within a single tree, we have to look more closely to discern those different niches, to perceive the capaciousness of life that a great valley oak supports. The nuthatch teaches this lesson as well as any bird, creeping down the trunk as no one else can, reaching into crevices no other bird can access. An oak is a smaller Earth, a world of branch and leaf overflowing with that same superabundant vitality and life. The bark is just its outer skin, but even hard brown bark is rich with gifts for the master mover on the trees.

CHAPTER 3
THE KINGLET AND THE GREENLET

Ruby-crowned Kinglet, *Corthylio calendula*

Hutton's Vireo, *Vireo huttoni*

A tiny bird hops through the canopy: gray-green above and paler gray below. I follow his every move with binoculars, straining upward until the familiar birding ailment known as "warbler neck" is begging me to stop. The bird's identity eludes me, and so I keep watching, gathering clues and details as I pursue this ceaselessly moving quarry.

The motion is what makes identification difficult—this bird never seems to hold still, constantly scuttling from branch to branch with restless, jittery movements. Even split-second pauses are animated by quick flicks of the wings, almost quicker than my eyes can catch. A hop, a hop, a quick flutter and hover to snatch at something on the underside of a leaf, a brief instant clinging sideways to pick at an opening bud of catkins. Among these constantly blurred frames, I struggle to extract details of photographic precision, the kinds of details that could be matched to definable field marks or the illustrations in books. Inconclusive glimpses patiently accumulate, fragments of a body dipping in and out of an ocean of leaves: a white wing bar, a ring around the eyes, the beak thin and small, and at last a faint but telling clue—a few feathers of red reveal the nearly hidden headpiece of a ruby-crowned kinglet.

Another tree, another bird. I am just a few steps farther down the path when I spot movement among the leaves. So many details seem the same: the dullest shade of green on the back, a pale gray on the belly, a comparable eye-ring, hints of white and black upon the wings. But something seems different.

First, I notice a difference in attitude: this second bird is not hurried and impatient, but calm and deliberate, each hop from branch to branch succeeded by a moment of careful surveying and consideration. The wing bars are not the same: two white lines surround a darker patch, while the first bird had a single white bar above one of black. Eventually she descends to eye level, some ten feet away from my silently watching figure. Raising my binoculars, I see her now as if perched before my eyes, fine details suddenly cohering into the bird one sees in field guides. Robust legs are coated in blue-gray, pale feathers form a wide bridge between the eyes, and a tiny, hooked tooth terminates the sturdy bill—a Hutton's vireo.

Many birders consider the primary point of interest of these two birds to be their visual similarity, a classic identification "problem" to be solved and set aside. Masters of modern birding like Rich Stallcup and Kenn Kaufman have penned careful guides describing the technical distinctions between the two eye-ringed leaf gleaners. And the puzzle of these birds' striking resemblance is indeed worth puzzling over, not simply as a subject in itself, but as an invitation and demand that we look and listen closer. Figuring out whether I am seeing a kinglet or a vireo requires me to look below the surface of my first impressions, to struggle through the limits of perception that conceal the constantly ongoing stories of the oaks. We typically notice such a small part of the world around us. One way to expand our circle of awareness is to seek the smallest and least distinctive birds that we can think of and watch, and listen, and watch some more, until their subtle differences proclaim themselves as clearly as words spoken in our mother tongue.

Ruby-crowned Kinglet: Belligerent Atom

Ruby-crowned kinglets are characterized in many observers' minds by two traits: smallness and restless belligerence. Conservative descriptions might simply note these birds as "constantly moving" or the like, but a more comprehensive evocation of the prime kingletian quality might be found in that family of informal words including "spunk," "moxie," and "yumph." Whatever you call it, kinglets have a lot of it. They are tiny but feisty, make up in motion what they lack in mass, and chatter and sing with a volume highly disproportionate to their size.

Compared to Hutton's vireos, ruby-crowned kinglets are the more well-known species, common and widespread across much of the North American continent in winter. While they nest in the forests of Canada and montane regions of the United States, ruby-crowns are extremely flexible as to their non-breeding habitat, spending their winters in forests, woodlands, scrub habitats, and residential neighborhoods without great apparent preference. Their small size, plain plumage, and refusal to hold still all make them less likely to be noticed and recognized by the casual observer than their abundance would suggest, but in many areas they are in fact a regular backyard bird and ready visitor to suet feeders.

In natural settings, kinglets tend to favor food that is commensurate with their smallness: most prey are under a quarter-inch in length, such as treehoppers, leafhoppers, plant lice, and scale insects. They pursue their minute quarry over and under the twigs and branches, moving through the oaks with energy and persistence as they identify one target after another; normal upright perching alternates with upside-down acrobatics and brief flutters to reach inside the convex leaves of the coast live oaks. The kinglets arrive in my latitudes in September and maintain their miniscule but manic presence until their departure for the north around April.

Smallness does not diminish personality. The opposite is true:

the kingletian character is practically defined by their extremely high yumph-to-bodyweight ratio. Of the other oak woodland birds, only bushtits and hummingbirds are consistently smaller, with ruby-crowned kinglets weighing in at around a quarter of an ounce. The miniscule bushtit weighs about as much as a nickel, a kinglet as much as a nickel and a dime, and the comparatively hulking Hutton's vireo tips the scales at a solid two nickels' worth of bird. Kinglets' lightness is essential to many of their classic foraging techniques, such as clinging to the most slender outer twigs and brief but frequent bouts of hovering to glean arthropods from the undersides of leaves.

At first glance, kinglets' constant motion may seem jittery and anxious, but the more you watch kinglets and see them interact with other birds, the more you realize that it is not fearful flightiness that lies behind their nervous demeanor, but energy eager for a fight. When ruby-crowns encounter a disturbance—whether human, competitive, or predatory—they react with immediate impetuosity, scolding, chasing, and revealing their usually invisible crowns. They are the avian equivalent of Jimmy Cagney's combative gangsters, diminutive figures bursting with pugnacious vitality.

Kinglets are plain and tiny birds, yet they stand out from all the other gleaners of the canopy, not so much in their appearance, but in their being—not how they look, but how they act. Kinglets are energy compressed, tight-packed with latent belligerence. The slightest spark may set them off. Then their hidden crowns burst forth and reflect the flame within them, always ready to erupt.

Hutton's Vireo: Demure Spirit of the Live Oaks

The birds of this chapter are not probers, long-billed reachers into hidden places, but gleaners, short-billed pluckers from surfaces. And so, while the probing nuthatch sought the deep-ridged bark of the deciduous oaks as an ideal foraging substrate, gleaners favor the perennial superabundance of leaves found on the smooth-barked but evergreen coast live oaks. Of no gleaner is this more true than of the Hutton's vireo, long known as "the spirit of the live oak tree."

There is perhaps no other tree more foundational to the experience of coastal Californians than the coast live oak, a tree readily identified by its stiff evergreen leaves, which are typically convex, unlobed, and pointed with sharp bristles (from which derives its scientific name, *Quercus agrifolia*, or sharp-leafed oak). I grew up among live oaks, went to school among live oaks, and lived for years with nothing but the thin canvas of a yurt between their branches and my body. My impression is that there is a great deal of truth in Donald Culross Peattie's claim that, "Whenever he can, the Californian prefers to make his home beneath an old Live Oak's shade." And so the intertwining of this bird's identity with these trees is for me central to their character. The live oaks welcome many birds, generalists like the kinglets graciously included. But Hutton's vireos are not passing guests. Hutton's vireos are the closest confidants of the always-living oaks, the birds who best know these trees, who know their secrets in all seasons.

Besides this oaken connection, the other preeminent vireo trait is unobtrusiveness and overlookability. Whereas kinglets are feisty and hyperactive, vireos are relatively demure and unexcited. Their quietness of demeanor enhances the smallness and lack of color that make both birds relatively invisible to the casual eye: Hutton's vireos, like kinglets, are a kind of pale gray on their head and breast and a dull greenish on their back. That green, far from pet-store parakeets though it is, provides the source for their names, both "vireo" (Latin for "I am green") and their folk name "greenlet."

Personally, I prefer common language names with suffixes of affectionate diminution and would be happy to revive this latter term: kinglets and greenlets both belong among the less-than-full-size things, linguistic companions to booklets, owlets, and piglets.

As for the rest of this particular greenlet's common name, I would propose a more informative prefix, such as "oak," "coast," or "Pacific," in lieu of "Hutton's," nomenclature that honors William Rich Hutton, an engineer and part-time bird collector who himself admitted discomfort with the name, given the slightness of his ornithological achievements. (This name—along with those of all eponymously named American birds—is slated for revision in the upcoming years, but its official replacement has yet to be determined as of this writing in early 2024.)

You may have never noticed them, but let me assure you: once you step among the live oaks, where the lace lichen hangs thick from the relatively moist and dense canopy of these evergreen woods, you are in the greenlet's home. Present all year round—they are the only nonmigratory vireo in North America—the oak greenlets maintain a very low profile outside of the nesting season, when they are prolific singers. I myself have often overlooked their presence, although I take some comfort in knowing that such neglect is not due to a particular shortcoming as a greenlet seeker on my part, but is primarily a consequence of the inherent nature of the bird. The old books are littered with phrases such as "not a bird likely to draw attention to himself," "distinguished by its demure ways," and "I realized that unless actually looked for they possibly would not have been noticed."

The Hutton's vireo is the demure spirit of the live oaks, the lichen-swaddled greenlet that moves calmly through the leaves and branches, pacific in more senses than one. Most people, even those who walk among the oaks each day, are unaware of their existence. But when you seek them out, here where the leaves never leave the branches bare, peering closely through the treetops and listening to every sound, then you may come to realize that the woods are more populated than you thought. It is not just the noisy titmouse knocking and the chattering of woodpeckers that give life and motion to the oaks. Someone else is threading softly through the stiff and convex leaves—the often silent greenlet conducting her wide-eyed survey of the trees.

How to Tell Them Apart

I recently spent a few weeks recording my every encounter with an eye-ringed leaf gleaner, making note of the crucial clues that revealed each bird's identity. I was left with a long list—there are many different potential indicators, and the most pertinent one varies considerably from one meeting to the next. If you only know one way of identifying each of these species, you will frequently be puzzled. If you know a dozen ways, you can cease from struggling to label a difficult-to-see bird, and instead enjoy your meeting with the miniscule fireball of winter or with the demure greenlet of the coast. If you know a dozen ways, then your friends among the oaks are shouting out their names.

Calls: That shouting is often literal. *Jit-jit-jit-jit-jit-jit. Jit. Jiddit.* Time after time, a kinglet shares this most common proclamation of identity, accounting in some form or other for fully half of the kinglet encounters in my ledger. Their endless *jit-jit-jit* notes are frequently projected in a long, continuous train, like a typewriter clacking away with jerky impatience in some bustling old newsroom. In addition to the generally sharp and projectile form of these distinctive sputters, kinglet calls are also often characterized by a doubled pattern, with the rapid couplet of a single *jiddit* enough to make the identification plain.

While relatively subdued compared to kinglets (what bird isn't?), greenlets still have plenty to say. As with most birds, this is especially true in spring, when males are singing, pairs are in close contact, and neighboring rivals engage in periodic squabbles. I set foot into a woodland dense with live oaks in early March and am immediately greeted by the greenlets' most distinctive call, the whinny—*heeeeeehehehehehe!* A longer first note slides then stutters, slamming on the brakes to skid across the pavement in a rapid series of whining yet staccato imprecations. Some sources describe this as a "laugh," but the typically elongated first note puts it clearly in "whinny" territory to my ears. Over the next hour, I hear the oak-horses on three separate occasions, as well as two instances of vireo couples communicating with gentle *whit* notes. These contact calls are softer, slower, less aggressive, and less frequently doubled than the harsh and insistent *jiddits* of kinglets.

Songs: *Z'wee, z'wee, z'wee, z'wee.* An endlessly repeated note rings out, slurring upward in a quasi-couplet over and over again. I walk on beside the creek bed, already dry after a less than torrential winter. Both banks are crowded by plants, now with poison oak, now with coyote bush, but continuously overhung with coast live oaks in this vireo paradise. A few minutes later a similar sound begins again, this time without the upward slur: *zooh, zooh, zooh, zooh*, simple, somewhat nasal toots repeated every second or two. Over the course of this same March walk, only a brief mile or so, I hear half a dozen vireos singing, in addition to the calling birds. I knew from the dominance of lichen-drenched live oaks along my route to expect good habitat for greenlets, but this abundance still surprises me. No one who visited these woods in the greenlet's unsinging months would have suspected this ubiquity. If you want to find the vireos, then there is no more reliable way than to walk into the live oaks in the early months of spring when their kazoos are ringing out from every seventh tree.

Kinglets, being winter birds in coastal California, and in much of the country, are largely not heard singing in my neighborhood—those rapid-fire *jiddits* are by far their dominant vocalization. But I must admit that on that early March day, a kinglet also sang, as they begin to do as spring develops, distant though they are from their breeding territories. This song is nothing like that of the vireo, and is in fact highly musical, varied, and complex: a series of high-pitched notes rapidly repeats to introduce the song, a variable second section follows, and then a third concluding section is often the loudest and most distinctive, consisting of rapidly alternated high and low pitches—*UPdownUPdownUPdownUP!*

Wing bars: I spot a silent bird moving in the treetops. Kinglet or vireo? The bird hops to an exposed perch, finally clear of obstructing leaves, and in that brief moment, I get a glimpse in profile, flanks and folded wings briefly frozen for my view. This is the clearest difference in plumage: vireos have two prominent white wing bars with a dark area *between* them, while kinglets have one prominent white wing bar in the lower position (plus a much fainter, shorter, and often hidden upper wing bar), with a black bar *below*. Hutton's have a wing-bar sandwich, with black between. Kinglets have one wing bar, with black below. Hutton! Sandwich! Remember those two words, and you will be ready when the wing bars flash before you. (There is one even more unmistakable plumage characteristic: If you see a ruby-red crown, then you are looking at a male kinglet. But the crown is usually hidden and so only occasionally helpful.)

Other visual clues: I meet another bird low and close, just a few feet above my head. Brief and fragmentary images come in and out of view as the bird moves through the dense foliage: a tip of a wing, part of a tail, legs and feet, a face—but no wing bars. Perhaps the next most useful field marks are the lighter lores (the feathered area between the beak and eye) of Hutton's vireos. Some people call this a "spectacled" pattern, with a pale bridge connecting the two light eye rings, but the bridge is so wide that it evokes ski goggles more than a typical pair of glasses. The bill itself can also be revealing; kinglet beaks are thin and straight, while vireo beaks are sturdier and have a tiny but distinct downward "tooth" at the tip. A final close-quarters field mark is the legs: kinglets have thin, fragile-looking black legs with yellow feet ("golden slippers"), while vireo legs are blue-gray and notably more robust in appearance.

Behavioral clues: Finally, if both birds are possible, given the season and the habitat, but the object of your encounter neither vocalizes nor presents a clearly revealing look, then your remaining evidence is in the bird's movement and behavior. As described above, behavior will eventually coalesce into quickly perceived impressions, but the most revealing qualities can also be rendered as an objective list of discrete characteristics. The overcaffeinated kinglets tend to dash about constantly, are always hard to follow, and give little nervous flicks of their wings at frequent intervals. They frequently hover to pluck at insects on the undersides of leaves and other hard-to-reach places. Vireos are comparatively sedate (although disturbance by a predator or rival can cause them to behave in a more kinglety manner) and tend to move methodically from perch to perch. Between hops, it is characteristic of vireos to look around, making several quick but unhurried turns of the head to survey their possible options from each perch. Greenlets are curious but calm, confident and assured.

The Double Spirit

There are many ways to recognize these birds, if you know their signs. I walk through the oak woods in spring and notice a dozen greenlets where others notice none: four singing, three whinnying, two calling, one flashing a wing-bar sandwich from the canopy, another coming down to investigate me with her hooked beak and ski goggles showing through the leaves, and one simply traveling through the branches with a placid, patient imperturbability unknown among kinglets. This experience states something true about noticing of birds in general: you see what you are looking for and hear the sounds for which you listen. The skill of identifying greenlets and kinglets is in the end the art of noticing things, made easier over time by knowledge and expectation. The skill of distinguishing vireos from kinglets is the skill of paying attention.

And once you start paying attention to these small and uncolorful, poorly known, easily overlooked birds, you will find that such attention is the key to far more than identification, the mere pinning of a name to something you see or hear. Knowing the whinnies and the ratchety-clack typewriter tunes not as facts, but as the voices of well-known friends, elevates your encounters from challenges of identification to opportunities for recognition and reunion. Knowing a kinglet from the briefest glance at the smallest part—a wing, a face, a foot—is not a mere end in itself, but the symptom of a deeper familiarity. When from a second's observation you can detect those differences in subtle motion that we might label as "anxious" or "assured," then you are not simply seeing motions, labelled with self-contained adjectives, and birds, labelled with merely identifying appellations. You are seeing the telltale behavior of particular characters, following along with the great story that is constantly flowing through the trees.

A day of April sunshine—the kinglets sing to announce their imminent departure and the greenlets sing to proclaim that they will not be moved. Green is everywhere, in the lush grass, the vibrant moss upon the trees, the pale lichens hanging from the dark leaves of the live oaks, and the bright new leaves emerging in company with sheaves of yellow catkins. And amid all this bursting exuberance of spring, two birds go about their business in the trees, immersed in the cascade of color.

The live oaks have long been my companions, the gray-trunked figures always standing in the background of my days. I ramble once again beneath their shade, seeking to get closer to the heart and essence of these trees, seeking the often silent greenlets who speak at last as spring draws forth their voices. The kinglets have not yet departed, but I no longer fret with neophyte confusion, fearing to mistake an imitator for the spirit of the oaks. Now I can ask and answer the deeper riddle of this duplicate: how can these unrelated twins exist, here together in the sharp-leafed trees?

There is ample room for both, because the true spirit of the live oaks is not singular and static, but multiple and multiplying. The always-living oaks shelter birds in every season; they offer food and safety when the wider world grows cold. And so in fall their spirit doubles with its mirror from the north. Now in spring that doubled spirit sings and the live oaks' life bursts forth.

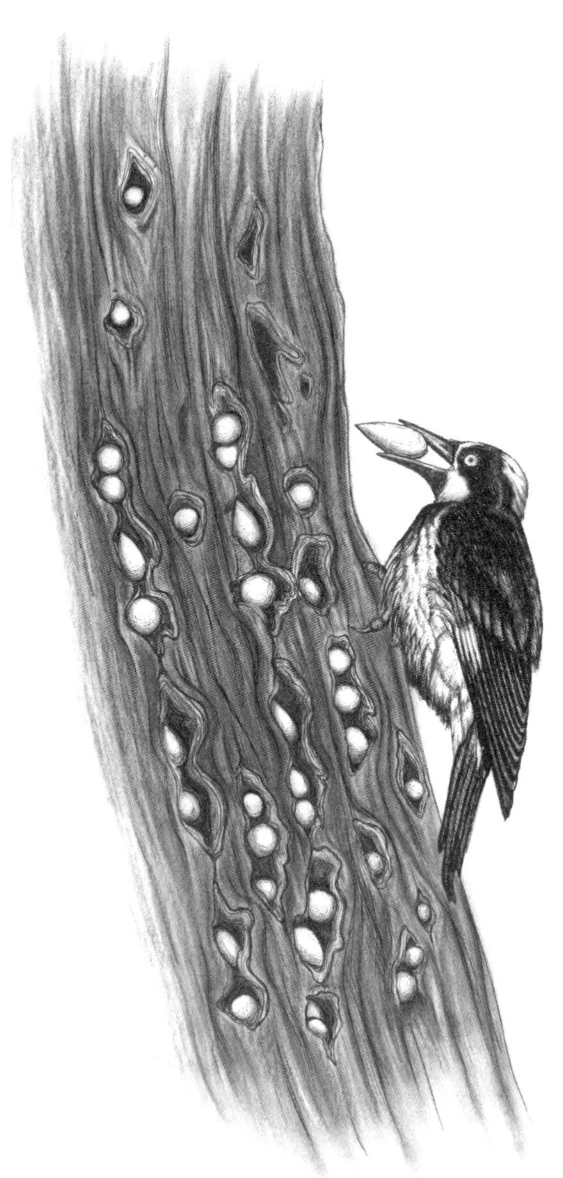

CHAPTER 4

THE RELATABILITY OF THE ABSURD

Acorn Woodpecker, *Melanerpes formicivorus*

WAAAK-a, WAAAK-a, WAAAK-a! A raucous burst of hoarse and overlapping calls erupts from the grove of valley oaks. The voices are not musical, like the flowing house wren song rising from the tangle of downed branches at the base of the trees. They utterly lack the gentle yearning of the mourning dove cooing softly in the distance. This closer cacophony is rough and grating, confused and disorderly, but also strikes me with an air of somewhat demented exuberance.

The authors of those sounds soon reveal themselves: acorn woodpeckers, a whole troop of chunky birds dressed predominately in black and white, both males and females capped in vivid red. In Spanish, they are sometimes known as the "carpintero arlequín," or harlequin carpenter, and in English they are sometimes described as "clown-faced," a similarly evocative shorthand. Large swathes of white and black divide the head, with bright eyes staring hugely from a black background; the net effect is generally suggestive of broadly applied, circus-appropriate face paint. But the description also works because these birds have a deeper clownishness, a pile-in-the-car disregard for rational calculation joined with a streak of exhibitionism.

I watch this group of five as they go about their business. Some fly in and out of the tree, making brief excursions to other perches within the grove or dashing out to grab an insect they've spotted somewhere in their territorial airspace. One bird chases another on a loop around the canopy, then both settle down to nod towards each other and exchange enthusiastic conversation, insistently urging each other to stop being so sleepyheaded (*WAKE-up! WAKE-up!*) or perhaps demanding their proper allotment of soft Italian cheese (*B'RRA-ta! B'RRA-ta!*). A jay approaches, hopping toward the base of one of their central trees, and two of the woodpeckers quickly swoop down upon the intruder with righteous, rowdy indignation, rather reminiscent to my mind of David Copperfield's Aunt Betsey chasing donkeys from her yard with a broomstick.

But the business of one bird in particular catches my eye, a female with a smaller red cap who climbs up a large dead limb, intent upon the surface of this old and bark-stripped wood. The panel is riddled with hundreds of finger-sized holes, some of which are empty and some of which are filled by acorns, one nut to each cavity. The woodpecker moves among her vast array of niches, spots a nut that's less than firmly lodged, and grabs it like a librarian spotting a misshelved book. She takes the acorn in her bill and hops in jerky, woodpecker fashion farther up the limb until a more appropriate niche is reached. She redeposits the nut, pointed end first, and hammers it securely into place.

I walk up to the tree, tracing the acorn-studded limb down to where the dead wood terminates at the height of my chest. I raise my hand to touch the acorns, solidly fixed like protruding nails driven into timber. I look up through the spreading canopy, the muscular branches extending across the sun-swept hillside. The clown-faced woodpeckers watch me from above, laughing to each other as I smile at their work.

Complicated Families

There is something unique about this scene: acorn woodpeckers live in a group, with the most complex social organization of any birds found in the oak woodland. Their system is sometimes described simply as "communal breeding," a phrase that somewhat blurs together two separate phenomena. The first is the delayed dispersal of young, with birds born in spring generally remaining with their family group throughout the winter (somewhat uncommon among birds) and then not infrequently *continuing* to stay with their parents as non-breeding "helpers" for up to five additional years (quite uncommon). The second aspect of their communalism is a truer polygamy, in which multiple males, females, or both breed and raise young cooperatively (extremely uncommon). About 9 percent of birds exhibit some form of communal breeding, but among those with helpless, nest-bound young, only five species in the world are known to actually discard the central touchstone of monogamy that is "knowing who your children are." The acorn woodpecker is one of them.

Let's start with the phenomenon of "helpers," the practice in which offspring remain with their parents and assist in subsequent breeding cycles, rather than dispersing immediately to form their own families. This behavior is not unheard of in the oak woodlands and in the wider world of birds—American crows regularly have helpers, and western bluebirds occasionally do, for instance. Choosing to help one's parents rather than striking out on one's own has several potential rationales: survival is easier (Mom and Dad help protect you and find food), you get more time to acquire useful know-how (especially in "smart" species that excel at experiential learning, like crows), helping your siblings feels inherently rewarding (not as good as raising your own kids, but better than nothing), and the housing market is tough out there (sometimes you need to wait for a suitable home to become available—or to inherit the family territory). This last factor seems to be especially important for acorn woodpeckers, a species in which prime territories with suitable nesting cavities and acorn granaries are often defended for generations by existing colonies.

The cobreeding, sometimes termed "cooperative polygamy" or "opportunistic polygynandry" is the more complex and unusual phenomenon, one which takes numerous different forms in acorn woodpeckers. In some cases, they behave as typical monogamous birds, with one male and one female occupying a territory and raising young together. Some groups, however, have multiple male breeders, some have multiple female breeders, and some have both. Including helpers, acorn woodpecker colonies of a dozen members are not uncommon. There is an important difference, however, between an acorn woodpecker commune and a human one: among the woodpeckers, members are all closely related in a tangled web that relies on long-term recognition of both parents and siblings to avoid incest. Cobreeding males are either brothers or father and son; cobreeding females are either sisters or mother and daughter. The tribes of acorn woodpeckers are not random assemblages of loose associates, but families.

It requires multi-year banding or genetic testing for us to read the exact family relations within an acorn woodpecker colony. But there are certain aspects of this life that anyone can see. We can see that acorn woodpeckers live in a group, sometimes of just a few birds, and sometimes of ten or more. We can see how they readily tolerate each other—the occasional game of tag notwithstanding—and often directly cooperate when chasing potential acorn thieves like squirrels, jays, or acorn woodpeckers from outside the family. If we watch during the nesting season, we can see several different birds entering the nest cavity, instead of the typical maximum of two. And in their great accumulation of acorns, we can see that this tribe of short-lived birds possesses a territory of greater age, a particular home that has been defended and expanded for generations: this group of woodpeckers is not merely one of the transient flocks that many birds form in winter.

But what we don't see are the personal stories of these birds, the relationships they have formed, ended, or resumed. We rarely can with birds, and so we often assume that they have no such personal narratives, no continuous tales of enduring connections. The most notable exceptions are birds of known long-term monogamy like titmice and scrub-jays, where we can observe a bonded pair and feel reasonably confident that a substantial relationship does in fact exist. Acorn woodpeckers, however, are the extremely rare case of a bird that maintains not just one bond, but many, retaining connections with parents and siblings year after year.

Imagine: A family may start with a pair of birds, much like any other—call them Mom and Dad. One day they meet, find a good cavity for nesting, and within it lay six eggs. Four of their children survive into adulthood and remain on the family territory. Their first fall sees six birds: Mom and Dad and their four children.

As the family's second spring approaches, several major developments occur. Two of the children move out after dozens of exploratory flights, travelling up to ten miles away in search of suitable situations in which to start or join a family of their own, leaving one son and one daughter still at home. A Cooper's hawk, however, disturbs the family tranquility and eats Mom on the cusp of spring nesting. Now is the season when young adults are on the prowl, however, and two unrelated sisters quickly move in from out of town—Stepmom 1 and Stepmom 2 have joined the family. With Oedipal risks now off the table, the remaining son graduates to become a cobreeder with his father. The Stepmoms Twain, however, see little desirability in keeping their stepdaughter around and send her packing, consigned to "floater" status until she can find a new home of her own. The second spring family status consists of four birds: Dad, Son, Stepmom 1, and Stepmom 2 (who will really be acting as Wife 1 and Wife 2 alongside Husband 1 and Husband 2).

With the family's increase in breeding adults—more bodies available to incubate eggs and find food for nestlings—the two females lay a combined total of nine eggs, from which six children survive to adulthood. No one knows exactly whose offspring they are; some are full siblings, some are half siblings, and some are combination cousins and half-aunts or half-uncles. The second fall family now has ten woodpeckers: Dad and Son, Stepmom 1 and Stepmom 2, and their six children of unknown specific parentage.

This process can keep on going, forming an ever more tangled, multigenerational web of close but non-incestuous relationships. There is one simple rule that keeps this working in a biologically sound manner: never mate with anyone of the opposite sex if they were present at your birth, although you may certainly cobreed alongside your same-sex parent or same-sex sibling. (Acorn woodpeckers do not *know* who their biological mother, father, or true siblings are, so they treat all the coparents of the family as potential relatives and therefore not to be mated with.)

The question of *why* acorn woodpeckers find it so uniquely advantageous to live continuously alongside their parents and their siblings has not been fully answered in all its intricacy and detail, but there is one great and striking correlation: the cooperative breeders are also the communal food hoarders. Acorn granaries are both naturally limited in abundance and require cooperation to create. When a good breeding territory is defined by its possession of a large, preexisting granary, along with all its associated oaks of diverse species, then a bigger family will provide more benefits (defending and enlarging the group savings) and the opportunities for independent breeding are fewer (the good granaries belong to established colonies).

These unusually extensive families are united by one unusual purpose: gathering, storing, and defending acorns, the great bounty of the oaks. When acorn woodpeckers live in other habitats where acorns do not dominate the available food resources, or in climates that demand migration, then their communalism diminishes or disappears. But when they live in California, surrounded not by one species of oak, but by two or three or four, then they see their safest way to make it through the winter: to gather food together and build something for their children.

Acorns can be saved in banks carved from oaks. These trees enable accumulation. Accumulation encourages cooperation. And the first and most natural form of cooperation is with those you've always known.

Their Life with the Oaks

There are many woodpeckers of the California oaks—downy, hairy, Nuttall's, Lewis's, pileated, flicker, sapsucker. But the acorn woodpecker is *the* woodpecker of the oaks. Unlike many of those other species, acorn woodpeckers do not venture out of range of the acorn-producing trees. And in California this relationship is strongest of all, with essentially all of our acorn woodpeckers choosing habitats containing at least two different oak species, whose acorns they store in those large collective granaries (in other, less oak-dominated parts of their range in the Southwest and Central America, the habit of acorn storage is less consistent than it is in California).

These granaries are unambiguously astonishing, and by far the most likely aspect of the acorn woodpeckers' lives to capture the attention of a casual passerby. Dawson reports and photographed a single tree with fifty thousand holes. One house on record from the Sierras was used as a bottomless granary, in which inserted acorns fell down into the wall cavity where they could not be retrieved—the house was estimated to contain more than sixty thousand nuts. Acorns are by far the most important food source for the species, with Pacific coast populations estimated to derive between 50 and 60 percent of their diet from these nuts of the oaks. Their importance is even greater than these numbers suggest when you consider seasonal patterns and necessities: spring and summer provide sap and insects, but acorns make up an increasing portion of these birds' diet in autumn and are the essential staple of winter. This dependence is made clear during years of complete acorn crop failure: without acorns, acorn woodpeckers generally cannot survive the winter on other food sources, and so they abandon their home territory, flying until they find a different stand of oaks that has the nuts they need.

Acorns are durable and can easily last through the winter, especially when they are stored above ground (where moisture, rot, and rodent consumption are less likely) and vigorously guarded. This is the basic act of prudent saving, evening out food availability through the changing seasons. One potential flaw in this strategy is that oaks tend to cyclically vary their acorn production, with superabundant mast years alternating with years of average production and periodic crashes. Weather conditions affect the timing of these cycles, so trees in the same area often have these crop failures simultaneously. Acorn woodpeckers reduce this danger by choosing areas with multiple oak species, particularly a mix of species with acorns that mature in one year (valley, blue, and coast live oak) and those with acorns that mature in two years (black oak, canyon live oak, and tanoak), thereby limiting the impacts of weather-linked cycles.

I've written elsewhere about scrub-jays, California's other great hoarder of acorns. Their strategy is different and relies on memory, with each bird burying some six thousand acorns each year, each in its own location. Such scatter hoarding is more vulnerable to forgetfulness (it has been estimated that jays recover only about half of their acorns), but those losses are an acceptable price for secrecy—jays are *hiding* their acorns from other jays, similar to nuthatches concealing their cached food. The cooperative strategy of acorn woodpeckers allows a greater accumulation, in safer environmental conditions as far as food preservation goes: they put their acorns in plain sight and *defend* them. They don't need to hide their stores, because they are working together. Suspicious individualists keep their money in cash, hidden in a back corner of the closet. But in a close and trusting family, such precautions are unnecessary: my wife and I share bank accounts and credit cards.

The strategy works: acorn woodpeckers are probably the most abundant woodpecker in California. Yet a common theme of the old bird writers is the seeming stupidity of individual woodpeckers in the execution of their caching. Numerous authors enjoy pointing out how woodpeckers will persist in foolish behaviors like dropping their acorns into inaccessible wall cavities, or storing non-food items such as pebbles, pine cone fragments, and chunks of bark. I've witnessed some poorly thought-through caching myself: I remember once watching an acorn woodpecker attempt to stuff a mouse carcass (I don't know where he got it) into a granary hole, with excess corpse components falling down to the ground in disarray, to the bird's confusion.

It's true and worth admitting: the woodpeckers don't exactly *know* what they are doing. They are not making the rational calculations that humans might make when deciding how much money to save or where to keep their retirement funds. But how consistently rational are we? Dawson follows up his description of acorn woodpeckers' endless and unthinking accumulation as "undoubtedly one of the most pathetic things in nature," with a comparison to a human miser, suggesting that the acorn woodpecker is an irrational and unwise creature for saving so excessively. But the possibility of this metaphor rests on a human reality: *we* are often irrational misers! If any creature struggles to pronounce their savings "enough," then I would think we humans would be the leading nominee.

So when I watch the clown-faced woodpeckers engaged in their endless pursuit of acorn accumulation, I don't disparage their ridiculousness, but embrace it as a phenomenon well-known to us in less benign forms, here translated into an acquisitive spectacle endearingly divorced from our more generalized greed or hunger for power. The woodpeckers aren't accumulating influence, money, or extravagant status symbols. They are collecting nuts.

Acorn woodpeckers are more like us than most birds among

the trees. They recognize their siblings and their parents, and continue to support each other throughout their lives. They are the original misers, with tendencies to misguided or excessive saving that are quite familiar to us humans. In everything they do, as in everything we do, there is an underlying ridiculousness, if you look for it—an extra-rational enthusiasm that makes us something other than simple calculators. Their savings and their household maintenance are a simpler species playing house. Their calls of *WAKE-up WAKE-up* are not serious mandates of diligence, but kids bouncing on your chest. They do the things we do, but with less pretension and self-seriousness.

Days spent among the oaks have many moments filled with beauty, from the strange and shocking stare of a spotted towhee in the shadows to the sudden burst of burning red when a flicker spreads her wings. But when I come to teach a child to love the world that flies among these trees, I will not preach of beauty or build a first foundation on abstract principles of science. We will look up at the clown-faced carpinteros and listen to their raucous laughter. We will find ourselves among an unshy crowd of brothers, sisters, parents. We will see their homes and labors carved from solid oak. And what I think will stand out in my child's mind and mine is the glorious kookiness and absurdity, the vibrant and ridiculous excess of the world.

CHAPTER 5
DISCONTENTED SHADOWS

Spotted Towhee, *Pipilo maculatus*

Leaves fall until they meet the soil, as do trees, and as do animals. But in that thin layer wait the seeds of every tree and shrub and flower that we see around us. From that skin of soil and soil-in-the-making, life ascends again. A tiny capsule sends roots down into the earth and a new tree reaches toward the sky. The oaks and bays, the buckeyes and madrones—all those slowly climbing columns rise from the dirt beneath our feet.

Some birds know the sun that touches the topmost spires of the canopy, where we are too heavy to follow. Others penetrate where we cannot, raising their children in beak-carved cavities of oak. But to learn all the secrets of the woods, one must know not just the birds that live *upon* the oaks—the titmice and nuthatches, vireos and woodpeckers—but also the birds *beneath* the oaks. And at their base and root is a bird that sifts through fallen leaves to find the seeds that start the forest, that sorts through fallen plants to find the life both latent and in motion.

This bird is called the spotted towhee. I remember when I first made her acquaintance, in the college botanical garden, remember my excitement when that bold composition of black and white, of rusty red upon the flanks and sunset red glaring in the eyes, would come briefly into view between the tangled branches. Now I know that what we see tells less than half the story.

Birds are small and wary, and live among the twisting thickets that obstruct our sight and movements. But they speak and tell us where to look, tell us who they are and what they're doing. In this chapter, I offer a portrait, not in blacks and whites and reds, but in three sounds that together paint a bird: a rustle in the fallen leaves, a discontented mewing from the shadows, and a ringing song that is compelled to climb up to the treetops. This book tells the stories of the birds among the oaks. And to hear a story you must *listen*.

The Rustle

Start with the humblest sound. Anyone might notice the grandest songs of birds or be struck by the overwhelming cascade of spring's dawn chorus. But if you only listen to highly electrified guitars, you might forget what it's like to pause and dive inside the reverberations of a single, clear-plucked string. The spotted towhee makes such a sound, one that's simple and inexpressive, and yet which says so much to those who know its meaning: the rustling they make while scratching through the leaves for food.

This rustling is a surprisingly attention-grabbing sound, produced by the singularly forceful two-footed kicks that towhees employ as they bat aside the leaf litter to uncover seeds and insects. The distinctive loudness of the sound derives in part from the towhees' preference for habitat with significant tree cover and abundant low, wind-obstructing branches, ideal for creating the thick, noisy layer of detritus they specialize in searching. But the unusual volume of their scratching is also a matter of technique: spotted towhees bring both feet quickly forward in the air, then rake them forcefully back through the litter, spewing scraps of leaves and twigs behind them. They often repeat the motion several times in quick succession, like a halting rope skipper who has to stop and reset after four or five jumps.

The objects sought by all this kicking vary by place and season, but include a wide variety of both plant and animal food. Almost innumerably assorted "weed seeds" are collectively the most significant category in their diet, but if a standout plant genus were to be sought in the lives of spotted towhees, oaks themselves would be the most likely candidate. In addition to being the most prominent source of both overhead cover and leaf litter throughout large swathes of the towhee range, oaks provide acorns, the most frequently consumed item in John Davis's study of spotted towhee foraging in California. Towhees also scratch for fallen fruits from such common oak woodland understory plants as coffeeberry, toyon, elderberry, and honeysuckle, as well as diverse beetles, ants, and litter-dwelling arthropods.

When I first started listening for birds, I would hear this scratching and wonder who it could be, suspecting that some excitingly large creature was just out of sight. This is a well-established towhee trick, not just a product of my own unique ignorance or overeager imagination. "He rustles the dry leaves like some animal twenty times his size," Forbush wrote of the eastern towhee a hundred years ago. "I often expected to find a woodchuck, or rabbit, or gray squirrel, when it was the ground-robin rustling the leaves," Thoreau concurred in the century before that, using an old name for the same bird.

But when it comes to the oak woodlands of California, I've gradually learned that there are really very few alternative candidates capable of making this strong and rhythmic shuffle. Only the large fox sparrows practice a comparable, two-footed forcefulness, but they are absent from my area from late spring through summer, and are sparser than spotted towhees even in winter. Most other sparrows are distinctly smaller, making them both less effective with this technique when in thick litter (and therefore less likely to employ it) and less noisy when they do attempt a round of double-kicks.

A useful comparison can be made with California towhees. This is a species that is just as large as its spotted relative and that sometimes double-scratches—but not one likely to cause auditory misidentification. This is in part a result of habitat preference: California towhees prefer more open areas than spotted towhees do, with fewer overhanging branches and consequently less leaf litter to rustle in. In the long run, however, those habitat preferences both indicate and shape physiological differences: litter-foraging spotted towhees have evolved to hop high and scratch hard. Both the proportions of their leg bones and the development of their leg musculature reflect these modes of moving and feeding. The loudness of a spotted towhee scratch is not merely a consequence of where a given bird happens to be scratching. Scratching is to a spotted towhee what hopping is to a rabbit or a kangaroo—their unique and long-honed specialty. These towhees scratch more loudly because they scratch more powerfully.

When you hear a towhee scratching, then, you're not just hearing *a* bird, some arbitrarily selected creature engaged for the moment in a noisy activity. What you're hearing is the result of both practice and evolution, nature's optimal solution to this specific situation. There is food hidden in the fallen leaves, stems, and scraps of bark. The most efficient way it can be located and consumed is by a bird of just the right size, with just the right seed-cracking beak, who has an unmatched ability to kick *just like this*.

On a relatively abstract, textbook level, this is a good example of resource partitioning, of the ways that different birds and other organisms develop adaptations to fill available ecological niches. But experiencing this in the real world is much more exciting than in a textbook: animals specialize because that is nature's way of filling every little corner of the world with as much life as possible! The world of fallen leaves is full of invisible seeds and countless crawling things, and these glaring, kicking creatures of black and white are the planet's extravagant response to that hidden abundance.

Many birds fear and flee the soil's surface, home to many dangers. Trees are their great refuge. But the red-eyed towhees have their walls of oaken trunks and branches, the concealment of dappled darkness to match their own dark and spotted cloaks. The towhees find their food where even other birds see nothing and lay their eggs in grass-lined hollows scraped from the root-bound blanket of the soil. No birds match their skill in drawing strength and sinew from this layer of detritus, deriving shelter for their children from the crushed and fallen fragments of the tall, unbending oaks.

The Moan

If forced to choose, I would name the spotted towhee's typical call as the most *expressive* vocalization of any bird in California. This expressiveness makes it easy to describe, and therefore easy to recognize: it has been compared to a cat-like mewing, defined technically as nasal and slurred (upward, most commonly), and evoked figuratively as whining or annoyed-sounding. The name "towhee" is derived from this call, albeit from the clearer-toned version of the eastern towhee, as are the southeastern folk name "joree" and the oldest recorded Native American name I can find, "chuwheeo." Such names are still applicable to the western spotted towhee, although our bird largely compresses the two syllables of "towhee" into one querulous croak, so a fresh and local transliteration would probably be something more like "rrrAAAHH."

Pete Dunne, a birder with an encyclopedic knowledge of bird vocalizations, describes this spotted towhee call as "a nasty, whiny, rising-and-falling, catbirdlike mew." "Nasty" is rather harsh to my ears, but the irresistible temptation to apply vibrantly biased adjectives is a testament to this sound's overflowing personality. There's no other bird sound upon which I could so easily riff in an ongoing flow of fitting epithets: it's an irritated groan, a querulous squeal, an irascible mew, a grumpy wheeze, a dissatisfied whine, a muted shriek, and a paranoid moan, all at once.

Such descriptions are more true of the nasal, slurred, and

somewhat intoxicatedly inarticulate voice of the spotted towhee than of the clear-voiced eastern towhee. Still, the close relatedness of the two birds is easy enough to recognize, and the distinction between the two species has often been questioned. As recently as the 1990s, the eastern and western birds were considered to belong to one species, the rufous-sided towhee. For my part, I often like to think of them within that joint identity, in part because the species boundary seems legitimately fuzzy, but also to discard the dry modern terms "spotted" and "eastern" and return to their classic joint moniker: the red-eyed towhee.

"Towhee" describes the sound they make, as is appropriate—I wish that more birds were named according to their voices, because it would remind us all to listen. But when I think of the appearance of these birds, I don't think of their feathers of black and white and rich red clay, but of those glaring eyes, peering out from the blackness of their face and the darkness of their home. I know no other songbird with such a stare, one that seems to belong to some plane beyond that of everyday mortal timidity. If I were the size of their typical prey, this shadowy demon might well appear like Blake's Leviathan, with eyes like "two globes of crimson fire, from which the sea fled away in clouds of smoke," or like his "mighty Devil folded in black clouds" who inscribes in corroding fires such pieces of pertinent wisdom as:

> Exuberance is Beauty.

And in the end, such fearsome portraits are not entirely divorced from the realities of this bird's life. Red-eyed towhees are not peaceable and innocuously social birds like goldfinches and bushtits. They may tolerate their mate's continued presence on their territory in winter, but they are not so closely tied to their partners as those amiable romantics the California towhees. Their black-set orbs of glaring red are not gentle, but they are exuberant. Their moans do not give off an air of friendliness, but they are meaningful, express volition, and suggest the existence of a more forceful personality than that of other birds.

This reading of the towhee groan may seem fanciful or shallowly anthropomorphic, but it has a significant scientific basis. In the most thorough study of the red-eyed towhee's vocal repertoire, Joan Beltz Roberts found that the mewing call comprises a range of meanings, from alertness to active antagonism, all falling under a loose heading of "disturbance." At the milder end, it can function as an attention-getting "look at me" call between paired birds. Calls that are longer, more quickly repeated, and harsher sounding indicate a greater degree of disturbance, such as that produced by the presence of a predator. As a whole, the extreme frequency of these calls seems to reflect the typical mental state of the spotted towhee: "general dissatisfaction with the current state of affairs."

Red-eyed towhees are frequently dissatisfied, and these calls are how they let the world know it. Other birds may experience a similar degree of disgruntlement, but few convey that sentiment so loudly and so often, and in tones that to our human ears remain reflective of that meaning. Roberts notes how the towhee's tendency toward frequent vocal signaling is not surprising in a bird that is so strongly associated with thickly obscuring vegetation. I would magnify this correlation further: towhees are voluble and vociferous *because* they are so constantly concealed.

Narrowly considered, their moaning indicates disturbance or dissatisfaction. But what does this moaning mean, considered from a broader perspective? In its frequency and volume, its expressive capaciousness and range of useful meaning, this vocalization testifies to the red-eyed towhee's whole way of life, hidden in the shadows. They moan to communicate across a realm they cannot see through, one crowded with branches and obstructed by leaves. Darkness conceals their bodies and their movements, and so the towhees' eyes can glower in the constant tree-wrought twilight. The ease of disappearance eliminates the dangers of detection, and so their voices rise and fall in unbounded discontent.

The Song

At the base of a favorite hillside near my home, the woods are dense and dark, with coast live oaks and bays thickly crowding around a small pond. A switchbacking trail climbs up its steep slope, travelling between clusters of oaks via grasslands dotted with coyote bush. As I ascend, the shifting topography grants preference to different trees, with black oaks and madrones joining the versatile live oaks.

And as I stand here in the middle of this hillside on a February morning, among the green grasses and shooting stars of early spring, I hear the red-eyed towhees singing from below and from above, sounding all around me on this hillside like a dozen different alarm clocks scattered across the slope. I stop to listen, trying to hear this song and everything it's saying, attempting to extract these birds' voices from the broader symphony around me.

CHWEEEEE! CHWEEEEE! CHWEEEEE! This song is neither elaborate nor musical, consisting in California of a rapid trill of buzzy notes, sometimes with a hint of an upward-swooping opening note. Each trill song is separated from the next by a short pause, perhaps six seconds or so. The rough and pitchless quality of the notes, the rapid rate of repetition within the trill, and the repetitiveness of the song as a whole bring to mind rather unflattering comparisons: some liken these songs to the buzzing of insects, others to the twanging of a taut rubber band. The songs of spotted towhees are not long, complex, or graceful.

Even the simplest of songs can be distinctive. Dark-eyed juncos and oak titmice are also singing this morning, both with songs that are technically trills, rapidly repeated series of the same note.

But they do not sound the same. Both juncos and titmice perform clearly slower trills than towhees, with individual notes countable or nearly so, in an objectively more musical tone, in which notes possess relatively discrete pitch compared to the unstaffable noise of the towhees. Towhee trills are fast and pitchless, until they hardly seem like trills at all in the common usage of the word: what they really do is buzz and rattle.

I continue listening to the endless towhee droning. And as these several singers carry on, I realize that the most important trait of this song is not contained within a single ringing rattle. I listen as I track the second hand completing one circuit around my watch: twenty-five towhee songs ring out over a minute's time, from perhaps three birds. The most important feature of spotted towhee song is volume and repetition, the quantity of songs one can cast into the air. Unmated spotted towhees will sing and sing and sing, trilling more than six hundred times per hour in the morning and accumulating some five thousand trills as a full day's work. I find giving a speech for an hour somewhat taxing and thirsty work. Shouting a single word every six seconds, six hundred times or more, would be considerably more challenging.

The pattern of this song can give us some clues as to its evolutionary function. In the case of some birds, song develops as a complex musical performance that demonstrates a male's health to potential mates. In others, a primary function is to provide a territorial warning to neighboring males, in which case demonstrations of higher cognitive function are less important. The spotted towhee song tends much more towards this latter category, though it should be recognized that "territoriality" is not entirely divorced from "mate attraction." In fact, one of the clearest trends in spotted towhee singing is a sharp decline in a male's song frequency when he finds a mate—that brief, bursting song should not properly be transcribed simply as "I-am-a-territory-holding-male!" but "I-am-a-territory-holding-male-currently-very-single-ladies-won't-you-come-and-join-me!”

As with the towhees’ rustles and moans, the nature of their song tells you how these birds live. Intensely repeated but simple songs indicate birds with a mate-attraction strategy based around highly audible displays of male territory holding. As with their querulous moans, these songs, unmusical though they are, contribute to a unique and vivid personality. Remember all the subtle distinctions that can be made if you listen, the insectoid buzz and the rubber-band twang, the endless manic shouts of *CHWEEEEE! CHWEEEEE! CHWEEEEE!*

The songs of all birds awaken you to the great yearly drama, the universal amorous upwelling that drives the spring chorus. Red-eyed towhees convey the essence of this season more forcefully than most: they keep on singing when many others tire, singing songs that strip out the inessential to repeat only the core and central feeling, that bottomless incompleteness that compels them to climb to the tallest point and launch song after song into the air.

A bird alone in spring is filled with discontent, is compelled by forces that neither prudence nor weariness can resist. And so the red-eyed demon climbs from his refuge in the shadows. He ascends through the twisting labyrinth of branches where he has so often groaned, securely hidden in obscurity. He mounts the oak before me, clinging to its topmost branch with nothing above him but the sky. He tilts back his head, opens his beak as wide as he knows how, and obeys that deep compulsion to buzz and burst and rattle to the furthest limits of his strength.

CHAPTER 6
THE GENTLENESS THAT BINDS

California Quail, *Callipepla californica*

A rustle—and I pause. This is not the strong scratching of a towhee, nor the soft and rapid skuttle of a fence lizard. I look around at the small sea of chamise bordered by oaks and see no birds except a scrub-jay silently watching from a treetop. The hills stretch for miles, with the wispy, leaning forms of ghost pines intermingling with the blue oaks scattered across the slopes. A red-tailed hawk slowly circles over the distant trees and utters her fierce and terrible call, softened by distance. But close at hand that rustle murmurs once again, now accompanied by a muted clucking, concealed and out of sight.

Put-put says a hidden bird, and I take a step toward the shrubs, craning my neck to peer into the dense screen of branches. In the first second of my motion the clucking accelerates, grows into a loud and frantic *put-put-put-put-put*, and in the second a small explosion erupts: a dozen previously invisible bodies burst heavily into the air like gray grapefruits with desperately whirring wings.

The quail sink to the ground like noisy balloons, their frantic flapping seeming only to slow but not reverse the demands of gravity after their initial brief burst skyward. Without pausing they touch the ground and immediately start running away down the unobstructed trail. Their feet are a blur as they speed into the distance, the several birds careening into synchronized retreat after the chaos of their haphazard flight, like marbles thrown upon a ramp that funnels their disparate trajectories with a newly harmonized motion.

Their escape was certainly effective—the quail are thirty yards along the trail and running quickly by the time I finish turning—but their demeanor inevitably has a touch of the comic, running with the ridiculous gait of humans with arms glued to their sides. This ludicrous impression is enhanced by their bottom-heavy, pear-like proportions and by the black, comma-shaped plumes that bob cartoonishly from the heads of the males. Even on the most ornate of human hats, feathers have generally pointed backwards in a mild nod to aerodynamism. But male quail have exceeded us in the evolution of impractical vanity.

That immediate visual impression is a large part of why people love quail: they are instantly and inherently amusing, endearingly ornate, and they vibrantly parody our own bipedalism, ridiculously inefficient in comparison to the effortless speed and agility of most birds. Fortunately for us, quail can be encountered across a wide swath of California and beyond, found not only in the oak woodland proper, but also in a wide range of neighboring habitats: in fern-filled forests, suburban backyards planted with diverse ornamentals, and in the true shrub communities of the coastal scrub and chaparral. In all of these places oaks are found, whether scattered throughout or on their heavily quail-frequented borders. From the perspective of the birds, however, the great unifying feature of their varied homes is the presence of low cover; when you're more a runner than a flyer, it is vital to have safe places at ground level to retreat to.

Safety and retreat—these themes play a large role in how we think of quail. Their general groundedness seems inherently vulnerable; they are protected by flight to a much more limited degree than most birds. But encounters such as these already point to the quail's remedies: skillful use of concealing habitat, yes, but also the collective efforts of a group that watches and warns when threats approach. Quail may not fly far from the ground. But they do not fly alone.

Finding Safety on the Ground

Most woodland birds are highly skilled at flying and possess only a vastly inferior ability to move at speed upon the ground. Winged escape is usually the first recourse of birds in danger. But there are a few birds who have found it useful to compromise their abilities of flight and enter our familiar domain of walking on two feet. Wild turkeys, introduced to California from the East, are one, the road-runners of the desert are another, and grouse up in the mountains and in the northern forests are a third. But the most representative of California's truly grounded birds are the quail: Gambel's quail in the desert, mountain quail at higher elevations, and the widespread California quail following the oaks up and down the state.

The walking birds have some broadly shared tendencies. One is the limited use of migration: while some montane species such as mountain quail make a shorter-distance altitudinal migration (marching down from high elevations in winter), long-distance latitudinal migration is essentially off the table for these little-winged walkers. The basic habitat need of low cover, primarily living shrubs but sometimes brush piles or downed trees, is also broadly shared across the ambulatory birds, with some variation. California quail avoid both uninterrupted grasslands and forests that lack such low brushy cover.

That basic requirement still leaves quail a lot of room to live in California, a land of diverse habitat and moderate year-round climate. Within areas with their requisite cover, they search for food on or near the ground, mostly consisting of seeds from low-growing grasses, forbs, and shrubs. Acorns, berries, and incidental insects round out their diet. Some seeds and fruit may be plucked directly from plants, while others are found by scratching in the dirt and duff. Larger than spotted towhees, quail employ a scratching technique that is similar to that of chickens, typically requiring only one foot at a time to yield effective results. In most of their everyday occupations, quail have no need to fly, and so, while they are capable of brief bursts of struggling flight, they rarely keep to the air longer than needed to escape an immediate threat.

If you are fortunate enough to have backyard quail, their feeding habits are easy to observe. I have lived in a few such quail-graced places, including the home of my suburban adolescence where I first began truly watching birds. Every day the quail would come, travelling from the surrounding remnants of natural habitat, grassy hills dotted with stands of live oaks and coyote bush. A quick flurry of wings took them up to the fence top, from where they would descend like unambitious users of a diving board, simply stepping off and dropping down.

Once over the fence, they typically congregated in the hedge, not from any threat in particular, but in accordance with a general principle of safety. From there, the flock gradually worked outward, clucking cautiously as their distance from cover grew, the bolder individuals venturing farther out to more exposed scratching spots. Searching under the bird feeder for fallen seeds was a favorite pastime, but a handful of scattered millet or a wide-basined birdbath could entice the whole troop to come high-stepping over the small lawn to wherever their rewards awaited. The opening of a door, a hawk's flight overhead, or the loud warning of a jay would spark a rapid dash to shelter, as notable for the silent, frozen tableau that resulted as for the blink-and-you-miss-it moment of actual motion.

The moments both before and after such alarms reveal the quail's finely tuned systems for safety. A first key factor in quail security is the promptness of the alarm, the conditions for which you can observe before danger ever appears: their flocks allow for an effective early warning system based on the use of sentinel males, rotating lookout assignments that typically leave a single male unengaged in foraging, often perched on some high point of a fence or shrub with no task other than watching for danger and sounding the alarm. When the larger winter coveys break up for breeding and one pair of adult birds takes on sole responsibility for their oblivious herd of pecking children, the male will lose measurable weight as he abstains for long hours from feeding, impelled to watch and warn.

Sociality provides warning—more eyes detect predators sooner. Quail's sociality is also a key function in the aftermath of an alarm. When all the flock or family have dashed into concealment, softly muttered *put-put* calls from those who can see the danger warn other birds to stay still and silent. If the threat proves passing, those outer birds will slowly venture back to feeding, relaxing their vocalizations to lower, rounder, and audibly less stressed notes to summon their hiding comrades back to business. If the alarm provoked a more dramatic scattering, then flock members will utter the California quail's most well-known call, the *cu-CA-cow* assembly call, which can vary from a gently muted, distant-sounding version to a loud and ringing rendition. This distinctive pattern forms the basis of many Native American names for the California quail, with recorded appellations including "se-kah-ke," "tah-kah-kah," "ca-ka-ka," or "kah-kah-tah." My midwestern father always hears them saying "Chi-CA-go!" but I prefer the more meaningful mnemonic "Where ARE you? Where ARE you?"

Sociality thus protects the flock through sentinels and sharing of information. Such measures go a long way toward reducing the risks of ground-based life. Social benefits might not seem, however, to alleviate one of the main sources of risk for non-arboreal birds: nest predation. Quail nest directly on the ground, lining a depression with grass, and such nests would seem inherently more vulnerable than nests up in a tree. This appears to be true especially in urban areas where a variety of mammalian nest predators thrive: rats, raccoons, and house cats can all be significant threats to nesting quail.

There are, however, counterbalancing benefits to nesting on the ground. While ground nests are more vulnerable to ground-based predators, in at least some natural habitats they are actually less vulnerable to jays and crows that tend to search for nests in trees. And so, in areas where mammalian nest predators are kept in check by their own larger predators—most notably coyotes—quail are more likely to thrive. And in terms of the overall danger of the nesting season, ground-nesting quail have a simple but important advantage over tree-nesting birds: they make the nesting period *shorter*.

Quail incubate their eggs for three weeks, at the end of which time their precocial young are immediately ready to depart from the nest, running after their parents as they toddle clumsily around and peck at what may or may not be food items. Significantly smaller arboreal songbirds might also incubate their eggs for three weeks, but then typically have another two or three weeks in which their young are helpless, often noisy, and confined to the nest with no possibility of escape. Young quail are, to be sure, still vulnerable, mobile though they are. But at least they can feed themselves, which greatly reduces the marginal burden of laying additional eggs: this enables quail to have quite large broods, regularly numbering more than ten birds, significantly more than the three to eight eggs per clutch typical of woodland songbirds.

The sooner that young quail can join the active family flock, with its watchers and warnings, the safer they will be. And the more young there are within each brood, the more likely some are to survive into adulthood. This early stage of abundant family life is clearly also the sweetest form of quail sociality from the human observer's perspective: precocial quail babies are endlessly endearing, fluffy little walnuts that obliviously pursue their enthusiasm for the world with minimal awareness of either external danger or their own limitations. They just run and eat and listen to their gently lovely mothers. Songbird childhood is brief and takes place mostly out of sight within the nests. With quails you can see a *family*, witness the well-known parental hassle and the uncorked energy of youth.

Ground-based life in all its chubby, comic running reads automatically as sympathetic. Quail's presence in both the woods and near our homes grants us a familiarity that we gratefully accept. Social behaviors of watching sentinels and varied conversation suggest gentle birds who work together. But I think it is this last feature of visible family life that most provokes our affection: I have yet to meet a person so hard-hearted as to be uncharmed by baby quail.

The Desire for Abundance

In the late 1800s, there were almost innumerable quail in California, more than I have ever seen:

> Quail in flocks now quite incredible soared out of almost every cactus patch, shook almost every hillside with the thunder of a thousand wings, trotted in strings along the roads, wheeled in platoons over the grassy slopes and burst from around almost every spring in a thousand curling lines.
>
> —Dwight Huntington, 1903, on the abundance of quail in 1875

It is unclear whether such abundance represented a historically "normal" quail population; Aldo Starker Leopold speculates that our many reports of quail superabundance from the latter part of the nineteenth century actually represented a short-lived quail "peak" stemming from the advent of small-scale agriculture and pioneer settlement, which broke up continuous forests and added hedges to what had been continuous grasslands. What we do know is that quail became noticeably less numerous over the course of several decades, with their previous abundance generally considered a thing of the past by 1925. Most contemporary observers of the collapse attribute it to overhunting, and we have abundant testimony of the overall volume of birds taken at the end of the nineteenth century: one hundred thousand quail were sold each year in San Francisco markets, a single market hunter could ship ten thousand quail per season, sport hunters could shoot two hundred a day, and commercial trappers captured hundreds of birds with each drop of the net.

To our credit, when the decline become apparent, the state took strong action in an attempt to reverse the trend. In 1880, the trapping of quail was made in illegal in California, and in 1901 the sale of quail was prohibited and twenty-five-bird bag limits were established. Throughout the 1920s, public-private quail refuges were established, growing to 1,200 separate reserves by 1931, the year in which the California quail was unanimously declared the official state bird by the legislature. At this time, the official Fish and Game policy was to minimize potential quail predators on sanctuaries and refuges, resulting, for example, in the killing of 684 house cats and 3,972 jays in the first six months of 1931. During the 1930s and early 1940s, 65,000 quail from game farms and Mexican imports were released into refuges.

The decline leveled off, but the past abundance did not return. While the ending of market hunting helped to stabilize quail populations, the reserves, attempted predator control, and introduction of captive-raised birds seem to have done very little. Leopold's overall conclusion is that habitat transformation—the replacement of native annual forbs by nonnative weeds and grasses, large-scale agriculture, declining water tables, overgrazing, and clearing of chaparral—was the overriding cause of the quail's diminished numbers. Leopold's solution—careful, ongoing stewardship of the land—is one that ultimately relies on local action and is difficult to achieve through any top-down, statewide imposition.

Although some of the historical interventions were of limited effectiveness, I find the overall scale of the policy response impressive. Many birds have negative population trends today, but that fact is generally considered a niche concern of environmentalists and conservationists, not an urgent societal issue that earns repeated high-profile legislation. The contrast is instructive: I think that lawmakers of the early twentieth century were able to take such significant action because a world of abundant quail was not seen as a goal in tension with human prosperity, but as a valuable ingredient of it. This particular vision of combined human and avian well-being enjoyed sufficient popular support because it was rooted in the personal experience of lawmakers and their constituents (including, yes, both hunters and eaters of quail).

I think that advocacy for birds is most effective and beneficial when it embraces those two truths: birds make human lives better, and those benefits must be *felt* before they can be appealed to for instrumental purposes. For me, the practical corollary of these maxims is that if we want to argue that birds are extremely valuable to human well-being, then we bird lovers should have self-evidently higher well-being than bird ignorers. To convince skeptics of the value of birds, that value must not be merely claimed, but also shown. If quail add to human well-being, then the defenders and appreciators of quail should not be sorrowful people, but the happiest.

Sometimes people ask me why I don't spend more time discussing the many threats to birds. I don't want to belabor the point, but I do want to avoid ambiguity and be forthright: I think that habitual negativity is not only excessively off-putting for public advocacy, but also directly harmful to the joy we take from birds and life. "Lamenting the troubles of others in the end helps no one," runs a sentiment attributed to Han-shan. "The clearest mark of wisdom," writes Montaigne, "is a constant joyfulness." The truth is that I pay attention to birds because they bring me happiness, I encourage paying attention to birds because I think it will bring others happiness, and I think recognition of our broadly shared self-interest will be better for the birds than more abstract moralizing that lacks this crucial grounding in personal delight.

To spread the recognition of threats and dangers in a way that is useful and practically impactful is therefore to spread recognition of the beautiful and comic, the relatable and endearing. The biggest problems of bird conservation stem not from sympathetic listeners who are inadequately informed of problems, but from a scarcity of truly concerned persons. And so I share stories of a past world rich with quail not to stir up useless guilt, but to inspire visions of a flourishing world that we could live in. I think that deep convictions require experience in support of logic, feelings of warm admiration at the sight of a strutting quail, memories of mornings spent tromping through the woods as calls of "Where ARE you?" ring out over the hills.

Male quail always make me laugh, their topknots bobbing as they trot out from the oaks. Young quail make me feel protective, as I watch them struggle through the head-high leaves. And I think that there are few birds more beautiful than a female California quail. The geometric elegance of the scales upon her belly are paired with the gentlest of faces, an embodiment of mildness and concern. Unlike the gaudy males, the females are plainly clothed in browns, pale and medium and dark, combining colors of comfort that my eyes read not as extravagant and strange, but as familiar and humane, as if the diverse mothers of the world all waited here to offer refuge to the fearful.

And so I watch and listen as her gentle, muted clucking summons her wayward-walnut children. A cautious handful of chicks emerge in response; the fields no longer churn with quail, frothing like the sea. But in the moment of our meeting, I forget the long-passed birds. Seven birds are here before me, and the earth cannot feel barren. Abundance is a fine and fitting goal, and this moment is so rich.

CHAPTER 7
A BIRD AT OUR LEVEL

Bewick's Wren, *Thryomanes bewickii*

A coarse, single-noted call of *whit, whit, whit* sounds repeatedly from somewhere deep in the impenetrable tangle of branches. A few years ago, a large and spreading live oak came crashing down, and now the broken limbs jut downward like barbed wire in a basic training obstacle course. I don't think I could crawl under there. But someone can. I still hear the bird, steadily *whit-whit*ing away as if mocking my inability to follow.

I have lifted my gaze up from the ground, above the layers of dirt and duff and fallen leaves where the quail and towhees scratch. Each bird has his realm and kingdom and that of the Bewick's wren is here, in the low labyrinth of shrubs and fallen trees, among the woody plants that don't grow too high and those that no longer grow at all. There are a hundred alleyways among the browns of bark and branch and old, dry leaves. The wrens are the ones who know the passages down to the hidden soil.

At last I catch a glimpse of the secretive bird, a tiny brown figure who darts up from behind a screen of jumbled twigs, only to immediately withdraw like a mouse into another of his myriad avenues of retreat. His movement is not one of hasty flight, but a confidently chosen route to his intended destination. Now he emerges again on a higher turret of his brush pile. He leaps to a perch atop the fallen trunk and stands revealed, far smaller than the skulking towhees, though by the volume of his chatter and buzzing one might have thought he was jay-sized at least.

A Bewick's wren is just a small, brown bird. They are not

mighty: they do not fly like falcons or hurtle through the air with the force of wild pigeons. They are not glamorous: wrens do not wear the finery of shining summer swallows or the brilliant gold of migrant warblers. The most prominent feature of their voice is not some ethereal fluting like that of the winter thrushes, but harsh, buzzing interjection as they patrol their humble domains with literally unelevated ambition.

I frequently wax purple and poetic about the vivid lives of birds, their visual beauty and their springtime symphony. But wrens inspire a different kind of warmth in me, incline me to instead wax prosaic and to celebrate the domestic. Life is not all grandiose soaring on the wind. Life is also going about one's business and making do with what one has. Bewick's wrens are objectively small and plain; they occupy a humble station. But in their voice and their demeanor, they inevitably give an impression of pride and self-possession—they are not humble in their minds.

Wrens Are the People's Birds

Many people have never heard of towhees, vireos, or kinglets. But almost everyone has at least heard of wrens. They are one of the most well-known of songbird families—arguably the most "classic" bird family in this book. Part of their wide repute is due simply to their geographic range, with wrens present around human habitations throughout both North America and nearly the entirety of Eurasia. The name "wren" has also entered popular awareness, however, because these birds have a recognizable and distinct personality that encourages relatability and lends itself to stories. On the one hand, wrens are unusually vivacious, often loud in their calls, striking in their songs, and rapid in their movements. But they are also plain and tiny, creating an endearing juxtaposition. One Chippewa name for wrens translates to "making a big noise for its size." Wrens are familiar but high spirited, humble in appearance but extravagant in behavior. The stories of this double character go back thousands of years.

Still, some may have heard of wrens without ever finely distinguishing them from their peers among the "little brown jobs." This term of birder cliché is particularly appropriate to this family: all wrens are quite little and most wrens are quite brown, with the occasional bold foray into grayness. Bewick's wrens run this limited gamut of color, being brown above and gray below, with plumage that is actually more boldly marked than the average wren, given their distinct white eyebrow line. Despite the family's chromatic plainness, wrens are easy to recognize, with a unique and memorable profile derived from one particular feature: their tails, which are routinely held in a characteristic cocked or upright position. Wrens point their tails up.

When surveying our historical views of wrens, one theme that quickly crops up is a kind of irrepressibility and pride, earning wrens the title of "king" from Europe to Japan. This designation goes way back—Aristotle mentioned the title in the second century BCE—and continues to this day in the German "zaunkönig" (hedge king) and Dutch "winterkoninkje" (little winter king). Unlike the ruby-crowned kinglets and their cousins, wrens' claim to royalty does not rest upon anything so superficial as a crown of shining feathers, but on their boisterous assumption of superiority to everyone they meet. There are few birds more quick to anger than a territorial male wren, no matter the size of the intruder.

One fable from Aesop demonstrates this sassiness. In the story, a wren and an eagle compete to see who can fly the highest. The mighty raptor takes to the sky, while the tiny-winged wren cunningly clings to the larger bird's back, only launching upward to claim the crown at the very moment when the eagle can fly no higher. A somewhat parallel episode appears in the Japanese folktale tradition: in this case, a little wren demands to join several hawks in their drinking party, but is challenged to catch a wild boar before he is allowed to participate. The wren improbably wins his battle by hopping into the ear of a sleeping boar, driving the massive animal to dash his head against a rock in his agitation. His mission accomplished, the wren swaggers back to claim his place at the raptorial punch bowl.

The arcs of both stories are the same: wrens refuse to admit themselves inferior to any other birds, no matter how large or how fierce, and in the end make good their claims, not through brute strength, but through a kind of impudent boldness. The net effect is not to argue that wrens *are* the kingliest or most majestic of birds, but to underline their weak flight abilities and general tininess—it is the discrepancy between their swagger and sauciness and their objective physical reality that is memorable.

This portrait is true. Wrens are, as a rule, bold, aggressive, and big talkers for their size. But they are also little birds that aren't expected to compete with hawks in hunting capability or altitude. That relative groundedness is arguably the family's most defining ecological trait, whether in the forest, marsh, or rock-strewn grasslands. For the widespread and generalist Bewick's wrens, their simple requirement of habitat is some kind of low, dense cover, whether in natural scrub communities, sufficiently scruffy yards, or the gently rolling hills of oaks. In studies of oak woodland songbird niches, the most distinctive feature of this wren's place is height: the great majority of their feeding takes place within ten feet of the ground. This layer includes both the lower branches of the oaks themselves, as well as common companion shrubs like manzanita or coyote brush in drier locations, huckleberry and salal in damp coastal iterations, and coffeeberry and toyon in mixed oak communities of moderate moisture. Wherever they find themselves, wrens—as the fable of the eagle illustrates—are not made to soar majestically over the earth.

In more recent wren culture, those core ecological truths of smallness, plainness, and groundedness are typical themes. Without the countervailing kingly gloss, these qualities are typically interpreted as cause for modesty. For instance, the English farthing for decades bore an image of a wren on the reverse—the smallest bird for the coin of smallest worth. Read about wrens in English between 1800 and 1950 and you will quickly come across the character of Jenny Wren, an archetypal figure of smallness, modesty, and earnest industriousness. The character had a wide influence: showing up in Mother Goose rhymes, serving as a nickname for a particularly petite and working-class character in Dickens's *Our Mutual Friend*, and granting an air of tidy thriftiness to the long defunct California grocery chain Jenny Wren Spotless Food Stores, for instance.

Such stories and metaphors are not our dominant modes of talking about wrens today. Tales of surpassing eagles and conquering boars are often considered quaint anthropomorphism. Characters like Jenny Wren can be critiqued as expressions of a subservient feminine ideal that merely co-opts wrens as symbols of the small and weak. Many might argue that what natural history content there is in these stories could easily be expressed in a more straightforward manner, without the errors or baggage of the past: if you want to say that wrens are small and don't fly high, then say that.

I am unconvinced. Sterile summaries of facts will never be as rich and vivid as knowing the stories of proud kings and modest Jennies. We need not hold old tales to a standard of comprehensive factual accuracy. Tolkien and C. S. Lewis once debated the limits of mythology, with Lewis initially maintaining that such stories were lies, "even though lies breathed through silver." But Tolkien won his friend over to another position, that "myth is invention about truth." From the high and ancient legends to the humblest child's rhyme, stories are not just isolated and fictitious narratives, but ways of making meaning from the real world we live in.

The stories we have inherited about wrens have their limitations and even inaccuracies, but they also have insights and advantages. In practice, memorable figurative truths will often be more valuable than dry facts that do not stick: if the only truths people are offered about birds are forgettable ones, then most people will forget them, and will live their lives apart from the irrelevant activities of birds. But when I see a wren come trooping out of his brush pile, beak buzzing and tail swishing at the presumption of a rival, then memories of the little king help shape mere sights and sounds into a story. And when I watch a wren gather fluff and feathers to warm the lining of her nest, I am warmed by recognition of a down-to-earth figure dressed in homespun brown, pursuing her modest domestic vision with thrift and tireless focus.

Bewick's is the People's Wren

A Bewick's wren comes regularly to my bird feeder, the easiest place to watch the species, given the family's proclivities for constant motion and impenetrable obscurity. She creeps up the slope, dashing from a clump of manzanita to a deer-shorn, scrubby live oak, disappearing for a moment before a dipping, furtive flight suddenly delivers her to the feeding station. Even here, she does not simply sit still, but approaches the block of suet with alertness and intention, takes a few determined stabs to fill her beak with the peanuty dough, and takes off, tucking her wings and tail in a descending flight back to her place of safety.

But for one suet-tempted moment, this wren stands revealed in both her wrenhood and her Bewickiness. Perched like this on top of the doughy cylinder in a proud conqueror's pose, I can easily see the most obvious field mark of the Bewick's wren, the bold white line above her eye. But even in the briefest glance, I could pick out the most obvious field mark of wrens in general, her cocked and upright tail. It is held vertically like a mop being carried through a crowded room, but swishes back and forth expressively like the tail of a cat or dog.

The specialist wrens have their special homes, many of them lightly traversed by people, if at all: the middle of a marsh, the depths of an ancient forest, a rocky patch in a wide expanse of unobstructed grassland. But Bewick's wrens are woodland wrens, and woodland wrens are more likely to be backyard wrens than are those other species. We build houses among the oaks when we can, and near the oaks when the prime real estate is exhausted. Even the most blank-slate of suburban developments or postage-stamp city yards will seek to emulate the woods in their small ways, adding shrubs and flowers among which wrens can weave. Bewick's is the wren that Californians will most often meet as neighbors, buzzing away beneath their windows and diving into hedges.

I find this sense of in-town neighborliness and commonplaceness fittingly evoked by this bird's namesake, Thomas Bewick, the great English bird artist of the late eighteenth and early nineteenth century (the name of both artist and bird is properly pronounced "Bew-ick," as in Buick cars). Bewick's wood engravings were praised by Wordsworth and Ruskin, by Charlotte Brontë and by Audubon, who named our wren in homage to the bird illustrator he most respected. Despite his cultured admirers, Bewick was overwhelmingly a *popular* artist. He was not a highly educated aesthete, but a working craftsman, a man who never had an art teacher, and one whose medium of wood engraving was ideally suited for inexpensive mass reproduction. When Audubon made a pilgrimage to Newcastle to visit Bewick in his old age, he spoke of him with reverential admiration: this bird was not named in a transactional act of recognition for a patron or procurer of specimens, but out of the deep admiration of one artist for another, the first to capture birds upon the page and share them with the world at large.

A Eurasian wren, as depicted in Bewick's *British Birds*

Despite their humble roots and decidedly unfancy-pants demeanor, Bewick's wrens, like Bewick, artist, loom large within their field. In spring, no day goes by when I don't hear their song, whether I'm walking in the woods or simply riding my bike through town. While the opening portion of that song can vary, it is easy to recognize by its ending, a clearly enunciated series of staccato notes on the same pitch. I see most field guides refer to this ending as a "trill," which is simply a very fast series of repeated notes, but compared to many of the birds I hear in California, the Bewick's wren sings a slow trill consisting of a relatively low number of notes, typically some six to ten. It is not like a faster junco's trill, thin and dry and fine. It is not like a much faster spotted towhee's trill, twanging like an unmusical elastic. The closing trills of Bewick's wrens are clear and piercing trumpet calls, often just slow enough to count. The other distinguishing feature of this "trill song" is that the trill is just the ending: there is always some kind of preface. Sometimes this will be just a few quick muttered intro notes, often there will be a buzzy note reminiscent of their aggressive disturbance calls, and in other parts of the country these introductory passages can be quite extended, variable, and complex.

These are the wrens I hear all through the woods and all through town: Bewick's are the most evolutionarily successful wrens in California, appearing in a multitude of habitats both natural and human-altered. If I had to suggest one discrete behavior that contributes to this adaptability, it would be this: their flexibility in nest sites. As with woodpeckers, nuthatches, and titmice, Bewick's wrens are cavity nesters. But whereas all of those other bird families are dedicated seekers or makers of cavities in trees, who can occasionally be tempted by a well-crafted bird house, Bewick's wrens are far more comfortable with unorthodox nest sites. Different wrens have different idiosyncrasies and generally recognized claims to fame. Bewick's wrens' claim to fame is their willingness to build nests in weird places. The everyman's wren nests everywhere: they do not discriminate with regard to housing type, dimensions, or neighborhood.

More or less natural sites in the record books include woodpecker holes, cavities in canyon walls, hollow spaces in rock piles and brush piles, niches under exposed tree roots, crannies behind bark scales, mounds in leafy forest floors, and perches on top of old phoebe and swallow nests. More or less human-aided sites include holes dug out of fence posts or burrowed into haystacks, an old shoe, a tin can, the pocket of a coat hanging on a wall, decrepit stovepipes, punctured sacks of seed corn, horses' feedbags, empty barrels and baskets, and open-entry mailboxes in active daily use. I've personally received a report of Bewick's wrens nesting for several consecutive years in an old motorcycle helmet. One study in West Virginia found that only 12 percent of nests were in classic tree cavities. In another study area in British Columbia, 40 percent of nests were in trees, but more were found in human structures, with a dedicated category required for "abandoned automobiles."

Some of these anecdotes may sound like old-fashioned artifacts of a more rural age. Today, many Bewick's wrens are found in comparatively tidy backyards, with nary a nosebag, sack of seed corn, or disused stovepipe to be seen. One remedy for this potential decrease in nesting sites is to put out a bird house, or nesting box, one of the sweeter and more benign ideas that humanity has ever stumbled upon. I have done my time in the bird store trenches, spreading the word about this simple opportunity for happiness both avian and human, and have counseled hundreds of people on nest box size, selection, and placement. When it comes to Bewick's wrens, however, it is delightfully hard to go wrong.

> Aspiring Landlord: Where should I put a bird house?
>
> Jack: Well, to have it be as attractive as possible to a variety of different species, I would put it six feet or higher and make sure it is solidly mounted rather than simply hanging loosely and swaying. Make sure you have a big enough entrance hole as well—anything smaller than 1 ½" is likely to exclude a lot of desirable tenants.
>
> Aspiring Landlord: But I had a house with a tiny entrance hole hanging loosely from a low branch three feet off the ground and Bewick's wrens nested in it last year.
>
> Jack: Hmm, yes. That sounds about right.

It is good and life-enhancing to have birds nesting around your home. I am warmed by the annual sight of birds singing, pairing, nesting, and feeding their young, by the enthusiastic continuation of this ancient cycle. I think a successful garden is one that's full of birds, one that imitates the oak woodlands in their diversity of plants, insects, and nesting sites. Few such gardens will be without a pair of wrens in spring. But if the artificial neatness of a manufactured landscape has left you wrenless, or only sporadically graced by our old companion among the smallest birds, then remedy your situation by the simple expedient of providing little boxes with little entrance holes in every place you can. Carry on the tradition; live in the company of wrens.

The little kings are all around us, down the street and through the woods. And so when I stroll among the oaks, seeking winged forms among the leaves above, I never cease to look down as well, down where the wrens are living, always at our level. Truth and Beauty are fine principles, but when it comes to the daily business of living, Jenny's virtues of modesty and diligence come in handy with an even greater frequency. I embrace the proletarian nature of Bewick's wren, named for the bird artist of the people, depicter of the everyday. And I am delighted each time I hear the story of an unexpected nest, glad to find another human eager for a world more full of birds. In my companion's laughing voice I hear that old tale once again, a tale of admiration for those who will not be told their place.

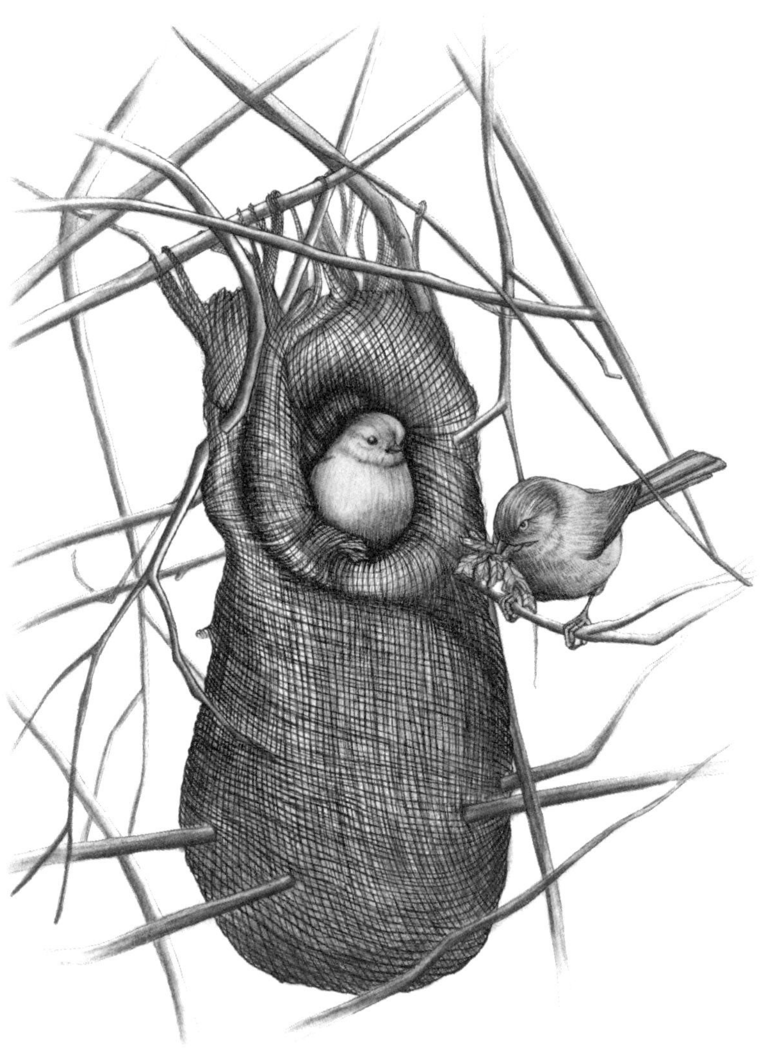

CHAPTER 8
THE LITTLE TAILOR

Bushtit, *Psaltriparus minimus*

Spending a lot of time with birds can sometimes make me feel heavy. Life as a human can easily seem ponderous and plodding; stairs and hills are just marginal intensifications of our constant weighed-down state. "Light as a feather," "free as a bird"—the standard clichés reflect the contrasting impression that birds give us of being seemingly unburdened by both physical heaviness and more abstract concerns, worries, and obligations.

Some sense of that lightness and freedom hovers in my mind even when observing some relatively earthbound species such as a quail or towhee. It starts to amplify as I watch the nuthatches and acorn woodpeckers climb determinedly skyward along the sunbound trunks of oaks. But even these birds seem anchored and tied down compared to others that rise still higher. Trees strain up and out, branches extend and grow finer, and twigs terminate at last in the fine filaments that hold the leaves that shake and rustle in the wind. And that is where only the most weightless birds can go: the vireos and kinglets, the warblers and chickadees, and most of all the bushtits, lightest of the light.

Twenty of these birds go careening through the woods with all the volatility of an unleashed basketful of ping-pong balls. I follow the spluttering of this constantly moving troop, picking out one pale-eyed female to watch as she goes about her daily business with characteristic impulsiveness. She darts from one twig to the next, unhesitatingly inverting to reach beneath a leaf with a few swift pecks of her tiny, stubby bill. The next perch spins her upright into a cluster of oak catkins; the following target sees her holding onto a stem with a single foot while the other fences with a recalcitrant petiole blocking her access to some invisible prey, hidden to my earthbound eyes by both smallness and location.

Bushtits can climb into the upper branches of the oaks, but they also tumble through the shrubs beneath with equal enthusiasm. What distinguishes their feeding niche is the smallness of the twigs they feed from: they avoid the trunks and larger branches where the nuthatches, woodpeckers, and titmice make their rounds. Bushtits are not like those birds, found at the heart of the oaks, but more akin to wrens in their proclivity for the dense and tangled, traits that often belong to plants of modest height. Like the wrens, bushtits can be found in many neighborhood hedges, or in naturally scrub-dominated chaparral. But oaks are not made of trunks and massive limbs alone, and their twigs and fine points must also be scoured by the smallest of all hunters.

The Smallest Birds

Of all the insubstantial and tiny birds that roam through the twigs and branches of the oaks, bushtits are the least substantial and the tiniest. A bushtit—bones, organs, muscle, fat, skin, and feathers—weighs about a fifth of an ounce, the same as a nickel. Bushtits weigh less than kinglets, half as much as chickadees, and a third as much as titmice. Their scientific name bears the specific epithet *minimus*, or "smallest," while affectionate terms of address found in the old books include "lowly midgets" and "dun-colored atoms." Bushtits are emphatically, strikingly little.

Being tiny is not just an arbitrary trait, but an ecological role. As a whole, smallness is the niche of all the arboreal insectivores; modest dimensions and light weight are prerequisites for accessing the outermost twigs, stems, and leaves, and for effectively foraging on tiny prey such as plant lice or scale insects. It is easy to overlook the profound difference in size between a *small* bird (a titmouse, say) and a *tiny* bird (such as a bushtit). But the difference is real, and has consequences for how these birds live their lives: one oak woodland study found that titmice forage 67 percent of the time on larger stems and branches, while bushtits spend 89 percent of their foraging time in the finer foliage. For a titmouse, midsize caterpillars are regular food items. For a bushtit, consuming such large prey would be like a human gulping down a jackrabbit.

Once we start to really look at bushtits, however, their intense smallness inevitably stands out as their immediately obvious characteristic. There are a few other details one might remark on if a bushtit would only hold still—the overall roundness of their form, their relatively long tail, or perhaps their eyes, beady black in males and light colored in females. But bushtits never hold still, and so our visual impressions remain imprecise, a blurry image of fuzzy little spheres darting about like a swarm of excited flies.

What do you get when you combine smallness, roundness, and plain plumage? Cuteness. Sianne Ngai, the leading contemporary theorist of the cute, has carefully laid out the constituent elements of this quality, in what could be a description of a bushtit. "Cute objects," Ngai explains, "have simple contours and little or no ornamentation or detail. The smaller and less formally articulated or more bloblike the object, the cuter it becomes. . . . The formal properties associated with cuteness [are] smallness, compactness, softness, simplicity, and pliancy." Bushtits *are* simply contoured, pliant-looking little blobs, or, as Florence Merriam Bailey summed them up more concisely in 1908, "plump fluffy tots."

Bushtit behavior reinforces this sense of cuteness. Besides their purely physical features, they give off a general air of harmlessness—their smallness largely prevents their acts of predation from being interpreted as such, with the rapid gulping down of a millimeter-long scale insect visually indistinguishable from a quick peck at an opening flower bud. Their strong sociability likewise seems the height of amiability to our eyes, with their constant lisping contact calls and eagerness to stay in touch suggesting a kind of clinging dependence equal to the most Velcro of dogs. You will often see a train of bushtits laboriously fly off, crossing some open expanse of space in a long, strung-out line. Belatedly, a final bushtit emerges from the now abandoned tree, spluttering with extra vigor and flapping her tiny wings with worried intensity, desperate to not be left behind.

Perhaps the most incontestable evidence of bushtit cuteness is one of their most distinctive behaviors: snuggling. One delightfully modest study noted how nightly observations of a bushtit roost saw a consistent practice of "clumped" roosting in cold weather: huddling together with other birds reduces heat loss, a danger that bushtits are particularly vulnerable to given their smallness and consequently high ratio of surface area to volume. But flock members have also been observed huddling in perfectly warm weather, and bushtit couples snuggle together in the nest as a matter of course (which is not a standard songbird practice). During the day, this behavior is most observable among recent fledglings and I have seen it on multiple occasions: a troop of clumsy youngsters pursue their parents as they beg for food, each following the sibling ahead of them to the same perch, where they seek to cram onto the branch in immediate shoulder-to-shoulder adjacency, a behavior which is not fled from or protested as it would be in most birds, but seemingly welcomed as a fraternal security blanket.

Many everyday bird lovers welcome and embrace cuteness when they encounter it in the avian world, just as the majority of human beings enjoy kittens and plump-cheeked babies. But serious birders often downplay it—cuteness is too easy to observe. Serious scientists often ignore it (although I'm not quite sure how you write a paper about huddling bushtits *without* remarking the adorability of the phenomenon). And scholarly thinkers like Ngai are apt to focus on the shortcomings of artificial, commoditized cuteness rather than resting content with the trait's more natural manifestations.

But I think ease of observation is a positive trait when looking for things to enjoy. Some aspects of visible cuteness tie into biological truths in a way that enriches our understanding. And in the end, cuteness is not a constructed phenomenon, but a natural reaction born from deeply evolved instincts. Cute creatures make us feel protective, nurturing, and affectionate. I think that fostering such emotions toward the world of birds is a good and life-enriching thing.

So embrace the cuteness of bushtits. Why limit your prompters of delight to babies and bunnies, kittens and puppies? Enjoy the plump fluffy tots for their plumpness and their fluffiness. Point them out to friends and passing strangers as an opportunity to laugh and smile.

"What are you looking at?" someone might inquire as you peer into the bushes, a line of six fledgling bushtits in your sights, huddled close for comfort. One has inserted herself backwards, facing the "wrong way" in the queue, ignoring the predominating orientation in her anxious cuddle-craving: five sun-dazed, staring faces look vaguely in your direction, with one isolated tail sticking out among the ludicrous assemblage. Why not share some happiness with those suffering under the misconception that cuteness comes from animators, ads, and toy shops? Let them know the truth as you share an earnest answer: "I'm looking at the cutest thing that I have ever seen."

The Friendly Birds

Following switchbacks on a steady ascent through the mixed evergreen forest, I stop, arrested by a sound. More accurately, I hear not a single sound, but many, a suddenly enveloping environment of vaguely located spluttering. Sharp, high-pitched *spit* calls seem to emanate from all the canopy at once in a staccato stutter. I peer into the trees and find the careening gray forms of bushtits, popping in and out of view among the foliage as if they were bobbing in water, momentarily on the surface only to disappear the next moment.

When I teach bird sounds, I try to focus on a key feature or two for each vocalization. Learn even one distinctive facet of a sound, something you can put into words and into your memory, and you will be able to recognize that song or call the next time you encounter it. For bushtits, the two most distinctive vocal traits are their most useful identifying traits as a whole: motion and multiplicity. You can hear them moving, and you can hear that they are *many*.

Their intense sociability is the most obvious behavioral characteristic of bushtits, whether detected by ear, eye, or meticulously collected data. Typical group sizes in coastal California are perhaps in the ten- to twenty-five-bird range, but some larger troops might include more than fifty birds. I remember one Christmas Bird Count when, prompted by the needs of the count and company, I attempted to tally the bushtits lurking in a thicket with more than everyday precision. We heard a few, saw a few, and put down an initial ten birds on our count sheet. But then they began to emerge in typical bushtit fashion, heading for some trees across the way not in one well-coordinated flock, but in a drawn-out string, one bird's sudden inspiration to head for greener pastures passing along the group, tipping birds out of the bush one after the other like hesitantly falling dominoes. "No, wait, make that eleven, twelve, thir—no, fourteen, fifteensixteenseventeeneighteen. Nineteen, twenty. Twenty-two, twenty-three, twenty-four-twenty-five. Twenty-six bushtits!"

The main exception to such regular flocking comes, as with many birds, in the nesting season. Bushtits nest in pairs like other passerines, but in numerous respects the common avian instinct toward monogamous isolation is relatively weak in this species. It is weaker spatially: pairs will defend the immediate area around the nest against stranger bushtits, but only a very small area, and even that with variable intensity. It is weaker in duration: once the young fledge, the whole family quickly reintegrates into a larger flock. And it is weaker in consistency: paired birds spend most of their time near their nest, but still travel more broadly within their flock's former foraging area, running into old flock mates with no apparent aggression. About all you can say is that during the nesting season it is *possible* to encounter bushtits in an atypically small group of two.

In some areas—rarely in California, but more often reported in Arizona and Central America—even the nest itself is not restricted to two birds, with "helpers" fairly common. Apparently, these helpers are not young birds delaying their dispersal from the family territory, as in the case of acorn woodpeckers, but are instead unrelated individuals hoping to either cobreed (in the case of females) or assist and then breed in a second nesting (in the case of males). These helpers are typically birds who had a failed breeding attempt nearby, seeking another opportunity at an already constructed nest.

In sum, these small bushtit "families" have neither the close relatedness nor the long-term continuance of acorn woodpecker groups, which have their solidly ensconced inheritance to bind the generations. They are temporary alliances, more selective versions of the larger flocks, in which two birds—or a few more—work together to build a nest and fledge their young. With that task completed, they return to the full assembly, a larger cohort of chattering conviviality that slowly shifts its membership over the course of each year as birds periodically move in or out of the group.

Bushtits are, in short, the friendliest birds of the oaks. They always want to be with others of their species, they very rarely fight, and they have only the vaguest notion of the term *trespassing* (the strongest legal concept in the eyes of most birds). They forbear entirely from song, the classic expression of avian egoism. Bushtits' amiability and constant eagerness for company combines well with their visual cuteness to form an almost ideally benevolent creature:

diminutive and fluffy, chatty and eager to please. But bushtits are not just fluff and charm. They have another talent that is not so apparent to the casual eye. Bushtits—those tiny tots of bouncing infantility—are the greatest architects of the woods.

The Tiny Tailors in the Trees

> Bush-Tits build the most beautiful bird houses . . . the most beautiful bird houses that there are in the world.
>
> —William Leon Dawson, *The Birds of California*

I remember the first bushtit nest I ever found. Like all bushtit nests, it was an extraordinary object: a foot-long hanging sock, carefully assembled through weeks of labor, with lichen, moss, and bits of leaves all firmly adhered with a thousand beakfuls of spider silk. It was both crafted and camouflaged, a work of architecture and a work of nature. I marveled at its fineness and detail, inevitably concluding that it was beyond our human skill to recreate such objects.

The nest was placed in a planted redwood on the Berkeley campus, suspended directly over a busy central walkway through which thousands of students passed every day. No one seemed to see the birds, and the bushtits for their part seemed nearly as oblivious to the human crowds that hurried or sauntered by. Diligent students kept their heads down, intent on the content of their engineering or art history courses, not realizing that a masterpiece of strength and insulation was hanging in front of them, that a work of natural materials far more intricate than any Andy Goldsworthy sculpture had been freely gifted to the school.

Bushtit nests are not like other nests. The typical songbird nest is a simple cup, designed primarily to contain the eggs and newborn birds and secondarily to conceal its contents from predators and insulate against extreme temperatures. But bushtit nests are much more ambitious, consisting not just of a basket, but also of a tube or cylinder some six to twelve inches in height that extends upwards from the bottom pouch. An appropriately sized entrance hole is constructed near the top of this sock, often with a slightly overhanging hood. This extended tube is far in excess of what is needed to contain the eggs and an incubating parent—compare the vaguely similar but comparatively ramshackle nests of orioles, which are usually under six inches in height, despite those birds being several times bigger than the minute bushtits.

So why do it? Why spend an average of four weeks—and sometimes up to seven—diligently working with your partner to construct this short-lived miracle? Because a bushtit's finely woven sock fulfills all the goals of a nest, and does so *better*. It doesn't just contain a few eggs and a single incubating parent—given bushtits' tendencies toward nocturnal huddling and sometimes cooperative breeding, their nests are built to have a substantially larger capacity. As many as fourteen birds have been found snuggled in a single nest, as cozy as my wife and I would be in our two-person sleeping bag if we also had a dozen infant children tucked in around us. A bushtit nest conceals the eggs better than an open-roofed nest would do, allowing parents to forage for an extended time without exposing them to view. And it insulates far better than a typical cup nest, wrapping the eggs in a warm and cozy layer of plant down or feathers (one counter reported over three hundred separate feathers in a nest lining) that similarly liberates the parents from the need for constant incubation. Nor are these benefits confined to cold weather: nests in full Arizona sunlight have been measured at temperatures of more than 110 degrees on their exterior surface, while remaining a comparatively cool 85 degrees inside.

In Spanish, the bushtit is known as the "sastrecillo," or little tailor, because this is their trade and talent. Two birds, one dark-eyed and one light-eyed, separate from the flock in February, looking for the perfect place to construct their joint creation. They search the oaks for spider webs, filling up their bills with the strong and sticky fibers, tugging and flapping to break the strands free. Together they deposit their mortar on a likely twig, piling it up until it accumulates securely. Next come the main materials, drawn from whatever plants are near at hand, but most often including lichen, moss, grass, and oak catkins, all bound together with more spider silk. They form a rim, and then a bowl, and then move upward, week after week adding material to their creation. They raise up the walls, dot the exterior with concealing decorations, and line the interior with feathers, fur, or plant down. At last, they do what many birds do not: move in together, huddle close, and enjoy the warmth that they have made through their unmatched skill and labor.

All nests are incredible, and no nests among the oaks are more incredible than these. They are incredible in the amount of focused work that they embody, thousands of individual fragments of diverse sizes and properties sought out and pieced together. They are incredible in their functionality, achieving all those ends of containment, comfort, and security with nothing but a mass of lichen, flowers, and spider silk artfully combined. And nests are incredible in their diverse and varied beauty, so many different shapes, drawn of so many different materials, no two quite alike.

But even under all these backstories and surfaces, the embedded labor, effectiveness, and beauty of these objects, the central phenomenon of nests never ceases to endear itself to me: they are things that parents build to cradle and protect their children. I find it warming to witness the everyday actions of human parents, the tiny gestures of comfort and support that mothers and fathers constantly offer in their solid ever-presence. Nests are an extension of that instinct made visible and concrete, care and concern holding the grass and catkins together as much as does the sticky silk of spiders.

So go out to the woods in March and April, listening for the little tailors and searching for their nests. When you hear them moving through the trees and see their weightless, pliant forms, do not dismiss them as merely cute and insubstantial. They are weaving something special among the live oaks' leaves. Feet and wings combine to collect mortar mixed by spiders. Bushtit bills gather the materials to frame, insulate, and adorn: lichen, feather, and flower compelled to fit the pattern that the smallest tailors trace. And when the sun goes down and night comes to the woods, cold air will blow among the branches and you and I shall shiver. Cold air will blow through young spring flowers, but down and feathers will enwrap the little tailors' sons and daughters.

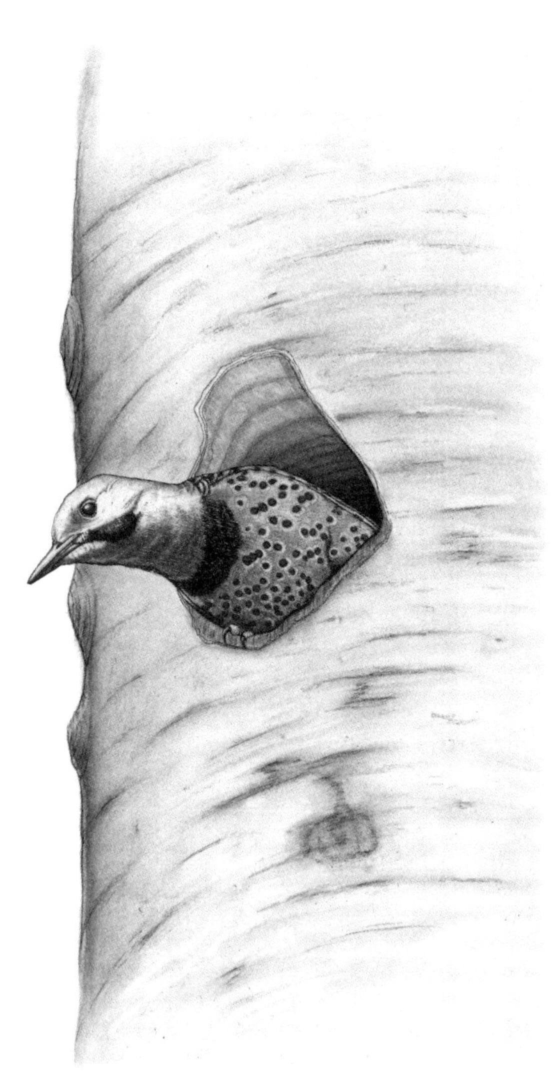

CHAPTER 9
A FIRE RISES

Northern Flicker, *Colaptes auratus*

One gray October afternoon, at that time of year when the California sun at last begins to show signs of weakening, I sat in a garden bordered by oaks. The birds of autumn had been steadily arriving over the previous weeks, and white-crowned sparrows now scratched in an untended bed alongside a resident California towhee. I watched them idly, temporarily perking up when a troop of waxwings flew overhead toward the more continuous woodlands where perhaps some madrone berries still lingered.

But then I noticed another bird down in the grass, a larger form than that of the sparrows and brown chippie. A gray-brown head rose briefly above the sea of green, then ducked back down into resumed invisibility. I watched the curious, intermittently reappearing figure as a few quick but somewhat awkward hops moved her through the grass and plants, a long and clumsy body of stripes and spots searching for something out of sight. She lifted her head once more, revealing a bill caked with moist soil, and then took off with a truncated chirrup and a flutter of lanky wings.

She flew away from me, flashing a white patch above her tail before climbing and circling to the top of the live oak, whose dark green leaves looked darker still on this overcast and sunless day. She rose up into the comparative brightness of the clouds, and for a moment she was frozen as she spread her wings and fanned her tail to land again upon the tree. And when she flew she was no longer a dull and spotted brown, and the sky was no longer gray.

What I saw instead was glowing reddish-orange, a burst of

flame that flickered with each beat of the bird's broad wings. The tail was more orange still, a vivid pumpkin color at the center of the fire. Her wings were shining like burnished copper, long feathers overlapping like plates of gleaming armor.

This bird is called the northern flicker, and is the most unusual of woodpeckers. The last four birds were birds of low height, birds of the scrub who come into the oak woods to inhabit the rich realm of their lower layers. The next four birds are birds of high places, birds of the tall conifers who travel south in winter or who represent great northern families, forest-birds among the oaks. Flickers are all of the above, transitional and border blurring. They are high nesters but low feeders, the most grounded of the forest birds. They are frequently found among the northern pines and firs, but they also live among the oaks, some year round and some when the nesting season ends.

The flicker feeds upon the ground and probes into the soil, but rises into the trees to dig out shelters for her young and those of other birds who follow. And when she flies, she flashes fiery copper, dashing living flames on the gray and wintering sky.

Woodpeckers Are the Builders

Woodpeckers are richly represented in North America, with twenty-two species present across the continent and at least eight found in California's oak woodlands. You know the basic pattern: tree-hammering, trunk-clinging birds of black and white: acorn woodpeckers fit this mold, as do the sapsuckers and the downy, hairy, Nuttall's, and pileated woodpeckers. For many, the trickier step is fitting flickers into that pattern, given their largely brownish plumage, dotted and striped in black, and their unusually terrestrial feeding habits. Flickers are unusual woodpeckers in some important respects, but they are woodpeckers nonetheless.

Take the archetypal woodpecker behavior: clinging to tree trunks and large limbs. All members of this family are characterized by their distinctive ability to move upon such vertical surfaces, refusing the normal concessions to gravity which incline most birds and other vertebrates to perch *on top* of branches. Many small and lightweight songbirds will cling from time to time, with varying degrees of comfort, but for woodpeckers this style of perching is their default and effortless preference.

The first major prerequisite for their clinging comfort is a unique foot structure. Woodpeckers possess what is known as a zygodactylic arrangement of toes, with two pointed forwards and two pointed backwards. This clinging enhancement contrasts with the evolutionary preference of most birds, in which three toes point forward and only one backward, which is better for gripping thin twigs. The second major adaptation that enables "walking" up and down tree trunks is the development of special reinforced tail feathers. On woodpeckers, the central tail feathers are stiff and strong, curving into a strongly flexed support when pressed against the tree, and are strongly muscled for careful manipulation. Flickers, like acorn woodpeckers and all other members of the family, possess both of these crucial adaptations.

The next distinctive ability of woodpeckers is their namesake: pecking wood. The fundamental strangeness of this bears a moment's reflection: woodpeckers make holes in trees by hitting them repeatedly with their heads. Imagine wanting to construct a small room out of a huge mass of solid wood. Imagine that you had no hands and no tools except for a large chisel attached to the front of your head, and that your method of excavation was to beat your head into the wood with all your force, over and over for two or three weeks until you had a cavity large enough to contain yourself and ten fully grown offspring.

Making this an effective lifestyle rather than a fruitless path of constant concussion requires a great deal of physical specialization.

The shape of woodpeckers' bills is one important element in the evolution of this ability, being formed like a woodworker's chisel, sharp at the tip, but relatively flat along the sturdily extended length. Equally important is the ability to withstand many repeated impacts, a capacity which entails numerous component adaptations: a longer lower mandible to transmit more force through the jaw rather than the skull; spongy, shock-absorbing bone at the base of the upper mandible; and skull development that protects the brain specifically from frontal impact.

There are good reasons for the evolution of this unique set of adaptations: pecking wood pays off. The most obvious practical use is in feeding. Many woodpeckers search for arthropod prey by hammering away at wood, including both particular concealing crevices that host hidden eggs and pupae, as well as larger expanses of dead wood filled with larvae and adults of boring insects. A secondary but highly audible use for woodpecker head bashing is "drumming," the family's equivalent to territorial song, in which a rapid patter of beak taps on resonant trees and other surfaces creates a loud and carrying sound. Flickers do drum on their nesting territories in spring, but are unusually inclined *away* from excavatory feeding—more on that shortly.

And then there is the third and final use for that chiseling talent, the one that I find most interesting of all: the excavation of nest cavities. Most woodland songbirds construct little cups of twigs and grasses. But woodpeckers have more solid nests, digging out chambers from the softer wood of dead and dying trees, or in living heartwood if necessary. These are the safest nests, with eggs well protected from harsh weather and invisible from the exterior. Even when the nest's existence is detected, the eggs are accessible only through narrow entrance holes that thwart most predators.

Cavity excavation is a very neat tactic, turning the solidity of trees into stout defenses for one's children. But what makes this ability of woodpeckers so ecologically important is that it creates safe nest sites not just for the woodpeckers themselves, but also for numerous other birds who use those cavities in subsequent years. Chickadees, titmice, nuthatches, wrens, bluebirds, swallows, owls, kestrels, and various ducks rely in large part on the work of woodpeckers (naturally formed cavities can also be used). Small "secondary cavity nesters" nest in the holes excavated by small woodpeckers. But for larger tenants, a larger woodpecker is needed.

That's where our bird comes in: flickers are the most important builders of bird homes in all of North America. Building a bird house is a wonderful thing: you make a safe little box, fix it to a tree trunk, and welcome more wrens or bluebirds to the world in spring. Flickers do this every year of their lives.

A male bird flies through the woods, until he finds material he can work with and a place where she'll feel safe. He pulls back his head to throw all his strength into that first blow of the spring, closing his eyes a fraction of a second before his chisel strikes the surface and the wood begins to splinter. For ten days he hammers: eight inches inward, and down a foot or more. His partner joins him in his labor, deepening the depression and laying seven eggs into a cradle lined with shaven wood. Soon more birds will join the world, emerging from this place of safety. They will not be the last to take their first flight from this shelter.

The Strange Groundedness of Flickers

Flickers are exceptional in the importance of their cavity excavation. But they are truly weird and unusual among our woodpeckers for a different trait: they find their food on the ground. The great majority of woodpeckers forage mostly on the trunks and limbs of trees, probing in crevices or hammering away at dead wood in search of invertebrates, which they supplement with various seeds and nuts.

What flickers do instead is to go down to the ground and eat ants. You will occasionally see them on the trunks of trees in the traditional fashion, and in the right time and place they are quite partial to berries, but over 50 percent of their diet consists of ants. Old stomach analysis studies found that approximately 70 percent of flicker stomachs contained at least some ants and 13 percent of stomachs contained *only* ants—including one packed with more than five thousand individuals. This formic focus sounds highly specialized, but in another sense it is an act of generalization—there are an estimated 20 quadrillion ants in the world, 2,500 times the global human population. Ants are everywhere.

Specializing in eating ants on the ground is not a limitation, but rather opens up a vast array of habitats that are more or less off limits to other members of the woodpecker family. You can go to both conifer forests and mature oak woodlands and find flickers in company with other woodpecker species. But you can also go to farmlands, scrub habitats with scattered trees, or suburban neighborhoods, and find flickers in those places too. This is especially true when the nesting season ends and birds no longer need those large, preferably dead trees for nest sites: many flickers are forest nesters, but they are not obligatory forest feeders. My town fits this pattern, with a smattering of breeding flickers in certain well-developed local woodlands exploding into abundance when autumn comes, bringing flickers from the mountains or the north in search of anting grounds not buried by snow.

This specialization in ground feeding encourages the development of other parallel adaptations. The most obvious is in plumage color. While typical black-and-white woodpeckers have good cryptic coloration for a world of mixed shade and sun, protected when predators look upward into the canopy or through a much interrupted landscape of foreground trees and background sky, flickers are mostly brown, with horizontal black bars forming a "ladder-backed" pattern that provides effective camouflage against a sunlit canvas of mixed soil and leaf litter. Flickers do have striking and exceptional colors, but they generally keep them hidden: only when birds flee the scene do their red wing linings flicker into life and their white rump patches jump into prominence, probably as an evolutionary tool to startle would-be predators.

Flickers' second adaptation for the ant-pursuing life is less immediately obvious than their atypical plumage: their extraordinary tongues. Woodpeckers on the whole have long tongues, often extending an inch or so beyond their bills. These are useful tools for pursuing insects lodged in trees; the extendable, sticky appendages reduce the amount of laborious demolition needed to reach prey. But flickers don't really want to excavate at all for food—at a glance, you might already suspect this by their slightly curved beaks, a less appropriate tool for daily chiseling. What flickers want to do is catch ants, often pursing them into tunnels or crevices. Their solution is a tongue that is not simply long, but extravagant, able to reach two and a half inches beyond the tip of the bill. Relative to the size of their heads, that's as if we humans could stick our tongues out ten inches from our mouths.

The important concept of convergent evolution describes how unrelated organisms develop similar traits in order to meet similar environmental demands. In the oak woodlands, for instance, woodpeckers and brown creepers independently developed similarly stiff tail feathers to support climbing on the trunks of trees. But the most memorable flicker-applicable example of this concept does not describe convergence merely with another bird, but with anteaters, those long-snouted, snake-tongued snufflers. So if you've ever found yourself marveling at a giant anteater in a zoo, as you watch that strange and alien appendage emerge from the long snout with seemingly independent volition, pause a moment and consider how a similarly long and sticky pursuer is constantly at work hounding ants all across North America and right outside your door. Anteaters are strange, outlandish creatures. Flickers are the anteaters of birds.

A clearing in the woods—grasses dotted with suncups and

vetch, patches of dirt seen at a distance—all this seems unremarkable at first glance. But a copper-clad, black-striped flicker is hopping along the ground, dipping her beak down to the soil. You pause and hesitate, making sense of those motions, imagining the details you are too distant to detect: that long tongue extending into a sea of ants, the scurrying masses climbing heedlessly aboard until that sticky serpent is withdrawn. The process repeats, the crowd thins, and the flicker continues her pursuit, reaching into each entrance of their subterranean caverns. The tongue inexorably floods a tunnel, pressing five ants against a wall. It pries into a crevice, extracting two more who had gone to ground. At last, you approach and the bird flies, red wings igniting as she disappears from view. You crouch upon the spot to find the ground still swarming, your attention newly directed to the often-ignored ants.

Predators come in all shapes and sizes—some grip and tear with talons and hooked bills, others growl and bare their teeth. But our world is full of oddities, creatures that don't fit our established preconceptions. We often think of hunting as an activity of brief and sudden efforts in the periodic pursuit of prey, each target carefully stalked and chosen. But for many birds, hunting is an activity more akin to gathering or grazing, a continuous collection of relatively unresisting insects. We might gather berries, employing all our agile fingers as we manipulate branches, fruit, and basket. A flicker gathers ants, plucking them at leisure with her strange and sticky tongue.

The Stories We Tell

> The red-shafted flicker is so well known . . . that any detailed description of its plumage or habits seems superfluous.
>
> —Willard Ayres Eliot, *Birds of the Pacific Coast*, 1923

Thus spoke Eliot a hundred years ago. And so it still should be: flickers are common, widespread, and distinctive birds. Most people have crossed paths with them, at least outside of highly urbanized areas. And yet I frequently encounter people who tell me they have never seen this bird, or who express disbelief that such a bird, large and glowing with red-orange wings, *could* be found within our latitudes.

This basic fact of ubiquitous familiarity is one of the most frequent themes in old accounts of this bird. "The Flicker is the most generally abundant and well known of all American woodpeckers," reads Forbush's classic *Birds of Massachusetts*. "The flicker, under whatever local pseudonym he may flutter, is one of the best diffused and most familiar of American birds," wrote William Leon Dawson in 1923. Even better evidence of the widespread flicker knowledge that once existed in the United States comes in the form of folk names, which are not mere testimonials to, but actual demonstrations of that familiarity: no American bird is more richly endowed with popular names. Over 160 different names have been recorded for this species, names derived from firsthand encounters by everyday human beings rather than ornithological authorities.

Folk names are embodiments of knowledge in words. They reveal awareness of the ecological features that I spent pages enumerating: flickers' proficiency at producing nesting cavities is reflected in names such as "high-hole" or "high-holer," while their feeding preferences are recognized in "antbird" and "ant woodchuck." Other vernacular names capture the audible calling cards I am always eager to evoke: "yarup" "yucker," "hittuck," and "wake-up" all respond to the two-syllabled *wick-a wick-a* vocalization that flickers make during courtship displays, for instance. In my area, where flickers are much more abundant in winter, I am far more likely to hear their most distinctive year-round call, a loud and resounding trumpet recognized in the name "clape" in imitation of the sound's strongly accented beginning, which rapidly falls and fades off. Even something as simple as the automatic recognition of those ringing *clape!* calls each time one names a bird embeds conviviality into life, especially in comparison to the default modern tendency to never notice those sounds at all.

Other flicker names are based on their visually distinctive traits. The black bib is acknowledged by "crescent bird" and the black belly spots by "spotted woodpecker." Others highlight the white rump patch: "cotton-tail," "silver dollar bird," and (creatively) "xebec," which Frank Burns glosses in his rich compilation of flicker names as "a small sea-going vessel carrying much canvas." But the dominant theme of the most common names is the brilliant color of the flicker's underwings and tail: this is, after all, what "flicker" itself refers to, as do the most popular alternative names of the eastern "yellow-shafted" form of the flicker, such as "yellowhammer" and "gold-winged woodpecker."

English-language folk names are, of course, relative newcomers to the world of flicker recognition: countless Native American names have doubtless existed across the birds' wide range. Burns helpfully compiles a dozen or so, along with literal translations. Some of these names are onomatopoeic in origin, but the majority refer to color, with frequent translations along the lines of "yellow woodpecker" or "golden woodpecker." In the Native traditions of California, where the red-shafted form of the bird is the resident species, glowing redness seems to similarly be the feature of note. Miwok stories tell of how Flicker sat close to the original fire, or how he once acted as "the kind one" who brought fire to the protagonist fawns of another tale. The most spectacular evidence of Miwok regard for that brilliant color is perhaps their ceremonial headdresses, each involving some 360 flicker tail feathers. These astounding creations are considered sacred objects that require careful handling and storage. They look the part. Such flicker feather headdresses are not a rare and merely local phenomenon, but have been found in most states of the West, with evidence of their use going back more than a thousand years. Flicker-red is one of the most wondrous colors in nature: those who were here first know it better than anyone.

We do not have any such universally shared culture of birds today. Not the quotidian and everyday knowledge of birds, such that we can talk to neighbors about flickers (or yellowhammers, high-holers, or antbirds) with certainty of being understood. Not the long tradition that imbued these feathers with a more than everyday glow across hundreds of Native communities. Too many of us lack connection to natural places and live our daily lives within our human bubbles, ignorant of such simple facts as whether the flickers of our home are winged in red or yellow.

But there is nothing preventing an individual from recovering something of that sense of culture, of knowledge passed down from past humans that still speaks of the birds we see today. The names, stories, and objects have not quite disappeared. Even if I cannot speak to everyone I meet about the high-holer flickering across the sky, the past can still speak to me, and tell me what it treasured in this bird of burning wings.

Today I step out in the autumn oaks, step out from the contemporary world of concrete and of cars, and see a flicker flashing red against the endless blue. A ringing trumpet echoes, a resonance that fades and vanishes like the lobed leaves fallen from the trees. Today the antbirds seek their prey, as they do in every state and province. Soon they will chisel out their homes, sculpting the nurseries that will soon host swallows and ducks, falcons and owls.

I see the orange flame glowing, as so many watchers have, and recognize that fire, woven into names and stories and stitched into sacred objects. To flicker is to show a light that vanishes and reappears. Bold paprika flashes against the sky like sparks blinking in the dark, until the bird flies out of sight and that red-orange flame departs. But the fire is not forgotten as I look out on the hills. All the stories are still with me, including those we write today. Flickers are etched into my memory and their fire always warms me.

CHAPTER 10
THE ANTIDOTE OF FEAR

Chestnut-backed Chickadee, *Poecile rufescens*

I perch cross-legged beneath a live oak, a little pot of oats cradled in my hands, and watch the chickadees. The morning is still young, the air still cool and fresh, and a small flock of small birds bustles in the foliage. A shade bigger than bushtits and a shade less frantic, the chickadees explore the undersides of leaves and the clusters of yellow catkins, snapping into place at the limits of the reaching limbs as if a magnet drew each bird in and locked her firmly in position. A steady gurgle of audible activity accompanies their motion, rapidly spluttered *tsick* calls and *tsick tsick* calls, nasal *dee* calls and *dee dee* calls, periodically developed into their namesake elaboration of *tsick tsick dee* when some particularly pungent point needs making.

Chickadee speech comprises one of a small number of "combinatory languages" known outside of humans, piecing together words like "chick" and "a" and "dee" in intentional assemblages of order and repetition to form distinct meanings. Our understanding of chickadee language, and all its specific dialects, is still very vague, consisting of scattered insights such as—in a rough paraphrase—*chick-dee-dee* means "great gray owl" and *chick-dee-dee-dee* means "saw-whet owl." Their vocabulary is limited and their syntax elementary, but chickadees have made the leap from words to phrases, and so I listen to their amiable chatter as if overseeing a room of foreign toddlers all poised upon the threshold of vocal intelligibility. No doubt some of those sounds mean something.

I watch one bird in particular, momentarily hunched and upright like a cat or fluffy gargoyle, with some item pinned to the branch between her feet in the classic parid style, a technique shared with the related titmice. Lacking binoculars, I can't tell what it is, whether arthropod or hard-shelled seed, but I see the chickadee repeatedly inserting her modest bill, tugging and prying and readjusting her hold as she dissects the object of her attention. The moment of relative stillness allows me to see her un-titmouse-like apparel; instead of the plain gray suit and pointed crest of her larger cousin, she wears an appealing combination of black on her cap and throat, white triangles across her cheeks, and rich redwood brown upon her back.

It all feels so companionable—the chickadees at their breakfast, me at mine. Many birds are understandably reticent, even standoffish, in human company, but chickadees are routinely cheerful and chatty, even in our presence. They offer the same companionship among the cool, dark conifers of the forest, interjecting welcome squeaks of amiability into the sometimes cloistral atmosphere of the creaking, wind-rocked giants. This flock drifts off, the second bird following the first, the third bird following the second, and the fourth bird following the third in a contented chain. I stand myself, bowl emptied and belly full, likewise replete with simple satisfaction.

All That Is Evergreen

Chestnut-backed chickadees are the second member of the family Paridae found in the oak woodlands around my home, along with oak titmice. In the titmouse chapter, I emphasized that bird's extremely strong association with the oaks themselves. Chestnut-backed chickadees, in contrast, maintain the bulk of their range in the coniferous forests of the Pacific coast. This difference is telling and points to the underlying biological and ecological differences between the two species. Chickadees are parids, but they are not primarily oak dwellers. Chestnut-backed chickadees are primarily the parids of the Douglas-firs and redwoods, and of the fir, cedar, and hemlock forests of the north. These chickadees among the oaks are on the outskirts of their world—I want to uncover where they come from, to see and feel the difference between a bird of the oaks and a bird of the firs and redwoods.

To distinguish the differences in character that stem from being a coniferous or oaken bird, it helps to first identify what these chickadees do share with oak titmice. The two birds both eat significant amounts of animal and vegetable food, have comparable suites of foraging techniques, engage in cavity nesting, and produce some similarly raspy vocalizations. The names "titmouse" and "chickadee" in some sense obscure their deeper relatedness—"chickadee" is an essentially American colloquialism, while within the broader Anglophone world members of both branches of the family are referred to simply as "tits." It is possible to think of the titmouse–chickadee division as making a simply visual distinction, with titmice being the crested parids and chickadees being the white-cheeked ones. My particular chickadee, the chestnut-backed, is a bird of relatively limited distribution along the Pacific coast, but a similar overlap of a chickadee and a titmouse is found in other parts of the United States, most notably with the tufted titmouse and black-capped chickadee present in much of the East.

But birds are not distributed according to the colors of their faces—some more significant differences must be in play to scatter chickadees throughout the firs and pines, while leaving titmice confined to the world of oaks. Titmice are larger overall than chickadees, leading them to eat larger food items on average, and also potentially to struggle to forage on the thinnest stems, including but not limited to those of conifers. Titmice also have a deeper, sturdier bill, which may help in consuming larger arthropod prey and which certainly helps in cracking and consuming acorns. Together, these traits seem to explain a large part of their distributional differences: oaks favor titmice, with their relatively strong, sturdy twigs and large acorns, relatively impervious to chickadees, while conifers favor chickadees, with their thin foliage difficult for titmice to navigate and their smaller seeds easy for chickadees to handle.

This is not mere happenstance of these two particular species, but the broad trend of their lineages in North America. It is not chestnut-backed chickadees that are uniquely northern and coniferous in disposition, but chickadees as a whole. They have winter in their blood. To see the several chickadee species of the West, I must go north and to the mountains. To see the titmice of the West, I must go south and to the desert.

Chestnut-backed chickadees and oak titmice do, however, overlap among the California oaks. They are *similar*, and their differences are correctly viewed as differences of relatively subtle degree, of comparative advantage rather than absolute capability. This is the nature of the distinctions that come into play as we explore the canopy insectivores in depth.

Watching titmice and chickadees among the oaks, I notice that many of their behaviors and physical locations are roughly the same. Both glean from the bark surface and from the underside of leaves; both cling to large limbs and to small twigs; both occasionally hover or catch a bug in midair; both hold hard seeds between their feet while opening them with their bills. Both, in short, are relative generalists, birds that eat a variety of both plant and animal food throughout the changing seasons of the year and throughout the changing habitats of their range.

An awareness of broader ranges and family traits, however, can give us insights that are harder to pick up from casual observations in a single location, during which these birds' similarities are apt to strike one more than their differences. Through carefully compiled, highly detailed data on their foraging preferences, we can just barely detect that chickadees and titmice live differently among the oaks: chickadees spend on average more time in the foliage and less time on the bark and solid branches than do titmice. Or we can glance at the family range maps and realize the applicability of a convenient shorthand generalization: this is a chickadee, and chickadees are forest birds, who live upon the fine and bending branches that shed the northern snow.

There is one more significant difference between the lives of chestnut-backed chickadees and oak titmice: chickadees flock in winter, living in pairs only during the breeding season. As discussed above, oak titmice are highly unusual among the Paridae in their conjugal isolationism, a fact I celebrated in the spirit of romantic enthusiasm. Chickadees are no nations of two, but gather into flocks comprised of both their own species and others, a fact I celebrate in a spirit of joyful conviviality and enthusiasm for abundance and diversity. The chestnut-backed passion for flocking is actually a component of their characteristic northerliness: winter flocking tendencies increase with colder temperatures, with the demand for higher food intake underlining the efficiency bonus of group foraging. In my winter live oak woodland, these flocks are the most exciting encounters of the season, centered around and announced by chickadees, but bringing a whole host of birds in their wake: vireos, kinglets, Townsend's warblers, juncos, nuthatches, and creepers.

The winter woods at times can seem deserted. Even the residents with year-round territories—like titmice, nuthatches, and wrens—are quieter and less combative. Sometimes I will walk for several minutes through the oaks without flowers and the trees devoid of song without a single sight of wings. But then a lisping, squeaking voice will speak from somewhere up above me and I look up in gratitude for the return of the birds' company. I would be glad to encounter a single living chickadee, but that is not what I meet. Instead I comb through the dull green leaves above me and find chickadee after chickadee, with further friends in tow. Juncos and kinglets, mainstays of the winter woods, suddenly materialize. A shy creeper moves upward on a trunk, a quiet presence I would not have noticed had the chickadees not alerted me to the continued existence of birds. And as I sort through the shaded branches, a black and gold vision briefly hovers as a Townsend's warbler brings a spark of light into the no-longer-quiet woods. Chickadees are the heart of the roaming flocks of winter, the central hubs in the wheels of life that spin through the cold and leafless branches of the black oaks and the buckeyes.

Chickadees are undeterred by winter, as Thoreau noted in colder Massachusetts conditions. They seem more alive than ever when the rest of the woods grow quiet:

> The fields are bleak. . . . The very earth is like a house shut up for the winter, and I go knocking about it in vain. But just then I heard a chickadee on a hemlock, and was inexpressibly cheered. . . . All that is evergreen in me revived at once.

Chickadees are forest birds come into the oaks, northern birds that reach toward the south. Awareness of their broader family provides a shortcut to understanding their behavior and distribution in the woods around my home. To bring awareness of broad principles and categories to bear upon individual instances is not crude generalization, but the ultimate goal of science. The books explain what I can see, and what I see vivifies and makes real the words and maps upon the page. Chickadees are evergreen and do not fall away in autumn, neither here nor in the north. Chickadees are evergreen and lead the flocks that warm the woods of winter and quicken them with life.

The Titmouse Dimension

Ralph Waldo Emerson wrote a poem about chickadees, a poem that contains all of my favorite out-of-fashion modes of enthusiasm for nature: anthropomorphic characterization of birds, a romantic elevation of our sensory experience as intrinsically important and ennobling, and an assertion that time spent in the woods can teach us how to be wiser, better people. I subscribe to the same old-fashioned school and concur: seeking the company of chickadees is good for you.

The poem's setting is a winter day in Massachusetts on which a lone walker is assailed by cold and fear:

> When piped a tiny voice hard by,
> Gay and polite, a cheerful cry,
> *Chic-chicadeedee!* saucy note
> Out of sound heart and merry throat.

Emerson can reasonably be regarded as a virtue ethicist, an eminent American proponent of the old Aristotelian tradition of identifying the character traits that constitute wisdom and lead to happiness. In this vein, the first virtue he finds in chickadees is cheerfulness. He is far from the only one to find their voice and behavior evocative of good-spirited sauciness. The friendliness of chickadees is a platitude of conventional modern bird education. They are less reluctant than almost any other songbirds to approach humans and are relatively readily enticed to feed on seeds or nuts offered by hand. Physical characteristics enhance this impression: their small round forms are utterly unthreatening and their squeaky *chick-a-dee* calls are entirely endearing. The human benefits of this are not abstract and theoretical, but parallel the most pragmatic and mundane advice for a wellness-enhancing social life: spend time with cheerful people. If everyone interacted with chickadees every day, I sincerely think the world would be more affable.

The second virtue Emerson identifies in chickadees is courage and stoutheartedness in the face of hardship and danger. The human narrator is cowed by the day's intense chill and envisions his death in the snowy night. But the chickadee expresses none of his dismay:

> Here was this atom in full breath,
> Hurling defiance at vast death;
> This scrap of valor just for play
> Fronts the north-wind in waistcoat gray,
> As if to shame my weak behavior.

In a figurative sense, chickadees are a vivid model of courage—their voice and behavior as we perceive them suggest this bird's unshakeable creed that "well the soul, if stout within, can arm impregnably the skin." I do not think, however, that this contrast between birds and humans is *entirely* figurative. There are entirely reasonable grounds to suspect that despondency in the midst of physical health is a human specialty, that chickadees are less likely to complain or give up than the average human faced with difficulty or discomfort.

A central passage of the poem proposes a theory of *why* chickadees possess these reservoirs of cheer and stoutheartedness:

> I think no virtue goes with size;
> The reason of all cowardice
> Is, that men are overgrown,
> And, to be valiant, must come down
> To the titmouse dimension.

On one level, this is simply a whimsical observation on the endearing smallness of the bird, whose cheerful courage is made more striking by his small proportions. But when viewed within Emerson's broader work of exhortation to courage and good spirits, I think that "the titmouse dimension" can reasonably be read as a proxy for the simpler life of nature, a life that shrinks and compresses not just its physical dimensions, but also its circle of concern, until the extraneous anxieties of overgrown men are discarded. "We interfere with the optimism of nature," he writes elsewhere. "We miscreate our own evils." To come down to the titmouse dimension is to shed the superfluities of civilization and live more in accord with our own nature.

What qualities, in sum, does Emerson's chickadee recommend to us? To be cheerful and stout-souled. What practices does the poem endorse? To undertake grateful pilgrimages to our little teachers and to live more of our lives "out of doors in the great woods." To visit chickadees is indeed to find "the antidote of fear," a counterbalance to numerous uniquely human manifestations of faltering self-assurance: exaggerated negativity, fearful conformity, clinging to comforts, and excessive intellectual abstraction that lacks physical grounding in the world of rock and tree and soil. The good life is not a topic to be considered indoors and then set aside when I step into the woods. The opposite is true: stepping into the woods is where I clear my head of cant and connect with real things. Time among the oaks simplifies and recalibrates our sense of good and bad, important and unimportant.

Pinning such thoughts on chickadees is not an arbitrary assignation of generic exhortations. Their amiability is real: chickadees are social, comparatively unafraid of people, chatty and vocal. Their boldness is real: chickadees are among the first to mob a threat or investigate a novel source of food. Their indomitability by winter is real: the range maps are not fanciful verse. Different birds, in the real and quantifiable specifics of their lives, summon up different reactions. The world of birds is a legitimate place to seek and find the various qualities we consider worth emulating. Chickadees are better teachers than despondency and doubt.

I grant that the poet adds something to his bird that is not strictly and literally biological in origin—such is the role of poems and stories. Cultures create myths, poets create poems, and we each create the stories that give meaning and coherence to our lives. As with any stories that we tell about those whom we admire—whether celebrated figures in the past or present world, the friends and family we know best, or the selves we wish to be—there is always an element of invention, of fiction that smooths the cracks in necessarily disparate lives. We long for coherent characters embodied in the world. Emerson created such a character with his chickadee, a portrait that has stayed with me for many years. And what he did with chickadees, I hope to help you do with titmice and towhees, flickers and swallows—to find the characters that can guide you to some wisdom, out of doors in the great woods.

Nothing clears my mind so well as day after day of walking, through many miles of trees. Distance erodes worries, hills massage out tensions, and footsteps foster daydreams. The heat of a long afternoon climb through sunny savannas gives way to the chill of the next morning, camped among the canyon oaks, trees that grip the rocks and scatter golden powder from their smooth-margined springtime leaves. The day begins with heartily squeaked whispers and cheerful gurgles, chickadees whose boisterousness somehow fits inside the quiet of the foggy, predawn shadows. The birds descend toward me, white cheeks puffing out against the blackness of their heads. The brown upon their shoulders is warm like varnished walnut, and eases tired eyes too used to glass and plastic.

The chickadees find me in a mood of similar benevolence, refreshed by nightjar music that washed away the weeks inside of doors. And in my chickadee-like cheer, I idly realize that they did not need my miles of exertion, my flight from roads and cities. They did not need to travel to the woods to find their better nature, for they were always here.

CHAPTER 11
THE WINTER THRUSHES

Hermit Thrush, *Catharus guttatus*

Varied Thrush, *Ixoreus naevius*

Each autumn, I see the first hermit thrush quietly appear among the oaks. She blends in, wrapped in russet, her cloudy breast thoroughly spotted like a leaf-strewn forest floor. Only her tail hints of color, but even that is an earthy umber, no vivid woodpecker red, but the tint of clay, nine parts brown to one part scarlet. She springs up to catch a berry, her wings blink, and she returns to earth, merging with the low tangle of branches and with the faded, fallen bark of weary madrones.

And when I walk a little deeper into the woods, when I come to the still dry creek bed where the live oaks and dark-leafed bays intertwine their leaning trunks and exclude the low, weak sunlight of the season, I discover another bird has appeared in my life: a small group of varied thrushes moves silently upon the cool, dark forest floor. At close quarters they reveal themselves as glowing birds, painted in charcoal and orange, carbon latent and combusting. Males contrast more vividly, like smoldering embers against an evening forest's darkness. Females are more muted, embers cooling at dawn as the earthen browns of the forest reappear. The folk name "golden robin" captures something of this impression, as does the name that arose in Danish, of all languages—they call this species "flammedrossel," the flame thrush.

But it isn't for their appearance that thrushes are chiefly known, but for their springtime songs. The hermit thrush is often considered the most beautiful singer on the continent, while the flame thrush may be the strangest and most haunting. Despite the many riches of California's oak woodlands, however, a profusion of nesting thrushes is not one of them—these two birds are only winter visitors, and you will seldom hear them sing here. Thrush songs are forest songs, chants of the firs and redwoods, while this is a book about the lighter and less lofty woodlands of the oaks.

But I have heard the hermit thrushes sing, heard the golden robins sound their foreign forest chords. And when I see them quietly emerge from among the winter oaks, I cannot erase that knowledge from my mind. I watch the winter thrushes and recognize something invisible but true: that they are not devoid of music, but have songs ripening inside them. I know the silent singers are more than they appear.

The Songs of Thrushes

There are around twenty different members of the thrush family that occur regularly in North America, including bluebirds and robins as well as birds formally known as thrushes. The family's distinctiveness of form is somewhat subtle, but not hard to recognize once you learn the pattern: small or medium-sized songbirds with full chests and an alert and upright posture. Excepting the bluebirds, most thrushes are inconspicuously-colored forest passerines that together form what Florence Merriam Bailey classically summarized as "the most remarkable group of sweet-voiced birds."

Hermit thrushes are often considered *the* sweetest-voiced of North American birds—a judgment that today combines objective evaluation of their song's euphony with a long-established culture of praise and admiration. The songs of varied thrushes are remarkable in a different way, consisting of isolated, strange, dissonant chords—strange, that is, in reference to our human standards, our notions of harmony and music. Both species share a certain tone common to many thrushes, a kind of whispered, fluting sound, a timbre that seems both penetrating and ethereal.

You can call it chance, a mere accident of evolution, but somehow this has happened: the songs that thrushes sing for thrushes are uttered in voices we are drawn to, whispers that penetrate our minds and reach into our feelings, that pass through our logical defenses and cast spells upon our reason.

The Hermit Thrush's Sacred Music

In the California summer, hermit thrushes live in the forests and sing from the treetops. They avoid the towns, and I make pilgrimages to hear them. Walk through the redwoods of the coast or the firs of the Sierras and listen: a pure, clear note pierces the air, followed by a soft, ethereal warble. Each phrase begins with that single high note, then concludes with a shifting and ungraspable, echoing and otherworldly phrase.

This is the most praised song in the history of American bird writing. Since at least the mid-nineteenth century, many of the most knowledgeable and articulate observers of nature on the continent have listened to hermit thrushes and concluded not simply that these are pleasant birds to listen to, but that their music evokes something deep and powerful, something higher and purer than we experience in our daily lives. And when listener after thoughtful, eloquent listener finds this to be true, I think we should ask ourselves what it was they heard, and if we can hear it too.

In the 1850s, Thoreau wrote about hermit thrushes again and again:

> This bird never fails to speak to me out of an ether purer than that I breathe, of immortal beauty and vigor. He

deepens the significance of all things seen in the light of his strain. He sings to make men take higher and truer views of things.

> [This song] changes all hours to an eternal morning. It banishes all trivialness.

A few decades later, John Burroughs recorded his thoughts on what he deemed "the finest sound in nature":

> [This song] seems to be the voice of that calm, sweet solemnity one attains to in his best moments. It realizes a peace and a deep, solemn joy that only the finest souls may know. . . . Listening to this strain on the lone mountain, with the full moon just rounded from the horizon, the pomp of your cities and the pride of your civilization seemed trivial and cheap.

In the early 1900s, William Leon Dawson added his praise:

> He who has not in his heart a separate place for the Hermit Thrush is no bird lover. He who has never heard the evening requiem of the Hermit has missed the choicest thing which Nature in California has to offer. . . . Mounted on the chancel of some low-crowned fir tree, the bird looks calmly at the setting sun, and slowly phrases his worship in such dulcet tones, exalted, pure, serene, as must haunt the corridors of memory forever after.

> One forgets all trivial things as he listens to the angelic requiem of the Hermit at eventide. Not Orpheus in all his glory could match that.

There is a defining antonym to which all three of these observers are drawn: the song of the hermit thrush is not *trivial*. Many moderns feel that it is a matter of basic scientific seriousness to set aside such old-fashioned talk of Orpheus and what "the finest souls" may know. But reading words from the past disrupts our merely contemporary assumptions: we have not achieved immunity from triviality. We are perhaps beset by it more than ever.

I remember reading Dawson's words for the first time, in a big leather chair in a quiet alcove of the university library. I read them and wanted to hear the sacred music he described. I gathered supplies—a hodgepodge of the cheap, the unfamiliar, and the makeshift—loaded a pack, and set off into the redwoods of the Santa Cruz mountains, seeking the songs that I had learned about in books. I was cold, sore shouldered, and uncertain; I had never backpacked alone before.

But I heard the hermit thrushes sing at sunset. The clear, high notes, effortlessly resonant, somehow free from the clinging self-consciousness that all human performances struggle to shed, followed each time by a swirling-echo fluting that hovered in the quietly darkening air. I lay in my sleeping bag and felt that I had found what I was seeking. I lay there as the sunlight faded, conceding without contestation the gap between the trivial and the essential, between life as we so often lead it and the life that sings at sunset among the darkly swaying trees.

The Varied Thrush's Haunting Strangeness

The ferns drip with cold water, the moss glows, and invisible dissonances ring out in the redwood-tethered fog. The harmony is strange, and there is no melody: each chord is enisled in a sea of resonant silence. There are voices chiming from the shadows, shadows that from time to time rustle with the suggestion of a hidden presence, but which refuse to tie actions to actors, to tie anything as mundane as a living body to these rain-plucked strings of the winter forest.

I see a movement. I stare in that direction, through the trees, across a slope, until I begin to decipher the murky forms moving through the dark woods of bays and live oaks grading into redwoods. A shape rises from the ground and disappears into the canopy, and then another, like bubbles rising silently through liquid to vanish at the surface.

At the moment that is all that I can see. Obscurity clings to these birds like perfume, and I see the winter robins as if through frosted glass. I hear them more clearly: isolated, resonant chimes sounding in the fog. Even among the winter oaks, there are days when you will hear the varied thrushes sing. The lack of melody, the seeming discontinuity between the widely spaced notes, focuses attention on the *tone* of each sound, which concentrates the ethereal fluting of the hermit thrush into a single echoing ring. The technically minded Nathan Pieplow usually defines bird sounds with words like “buzzy” or “upslurred.” How does he describe this voice? “Haunting.”

What is stranger still about this song is that varied thrushes often sing in not just one voice, but two, generated independently from their left and right lungs to produce a chord of inhuman harmony. We are used to the gentle consonance of pianos and guitars, approved notes combined in approved ways. But the flame thrushes don’t care about our rules. To emulate their harmony, I would have to take a pair of violinists, direct them each to play an arbitrary and uncoordinated pitch, then reposition their fingers on the unfretted strings like an exacting but tone-deaf teacher ensuring that every note is out of tune.

The song of the winter robin is simple, but still musical: a gong, a chime, a touch on the strings of a harp. And when that day comes when you arrive in the right piece of forest at the right moment, when the right confluence of atmospheric portents strikes a whole haunting of thrushes into one mind and feeling, then you will hear the trees ring not with one touch of the strings, but with an enveloping cascade, as if your ear were bent low to an undamped piano of a hundred mistuned keys. All the while you will stare skyward into the backlit, clouded canopy, seeking the source of those unearthly echoes, and seeing only shadows.

What the Winter Thrushes Teach

When you encounter these birds among the winter oaks, your knowledge of their summer songs does not fade into irrelevance. That knowledge means that you have grown in your understanding of birds and are now aware of things beneath the surface. Considering the traits you notice in December to be consistent and unchanging aspects of a thrush's character, when they are in reality traits of winter only, would be a mistake. Recognizing the *contrasts* presented by birds in winter is essential to understanding the full picture of their nature.

THEY DISPEL IGNORANCE OF SPACE

Spring birds are yet to appear, though the world of plants is already laying the groundwork for their arrival: buckeyes hold their green candles aloft, fists poised to unclench, while bays prepare their flowers and minute white bells fallen from the manzanitas litter the path like snow. The air is cold, few insects fly, and neither do many birds. But varied thrushes explore the leaf litter, quietly shoveling leaves and twigs aside with their beaks to uncover what lies hidden, from arthropods to acorns. Six months ago these woods were different: the varied thrushes were not here. Now they are.

This is the first and simplest lesson of winter birds: they travel, coming from other places beyond our own, usually stationary, home. Many people lose track of this, knowing it in their minds, but not living it with their own eyes and ears. It's easy to miss the June appearance of the first shorebirds, the July arrival of the great flocks of sandpipers, the August entry of the dabbling ducks, and the September flood of songbirds: the autumn songs of crowned sparrows, the passing wave of warblers, and the quiet, subtle infiltration of the winter thrushes. The world is richer than we perceive.

Birds have good reason to come south. Most of the varied and hermit thrushes that winter in coastal California breed somewhere between Alaska and British Columbia (a smaller portion of the varied thrushes nest as far south as northernmost California). But for all but the hardiest birds, Alaska is no place to be in winter, with fruit and insect prey very hard to find. Food availability is the first consideration that brings these thrushes down from the northern forests.

Watching birds is, in large part, watching birds eat. Noticing where and how birds search for food tells you a lot about what kind of birds they are, clues you in to where they fit in the grand patterns of the seasons and the continent. Both varied and hermit thrushes enjoy berries, especially during the first months after their autumn arrival. Berries tend to decrease in abundance by midwinter, however, and the thrushes then shift their focus elsewhere. Varied thrushes continue to eat a mix of durable vegetable food (acorns and small seeds), while consuming an increasing proportion of leaf litter creepy-crawlies including insects, pill bugs, millipedes, worms, and snails. The arthropod prey of hermit thrushes is somewhat less ground focused, with their light weight allowing easier hovering and gleaning from leaves, but the same generality is true: when berries are depleted, thrushes feed most often on the ground.

The winter thrushes feed most often on the ground—this is not surprising. Many prominent ground-feeding birds like flickers, crowned sparrows, and fox sparrows appear here around the same time, while aerial flycatchers and swallows largely depart. Summer is the time for sky birds, but in winter, the air is less alive and the food is on the ground. You can see this in their very feathers, browns and dark tones replacing lighter whites and grays and yellows. In winter, the ground-colored birds come from the north and the sky-colored birds head south.

I watch the winter thrushes, gray and brown forms upon the brown of the forest floor. They have come from somewhere far away, from beyond the borders of even this huge country that I live in. I watch them, and see that I live within a larger story. A hundred million birds flew to the south, in search of warmer skies. A hundred million birds flew from the north, finding all they needed in the fallen leaves beneath my feet.

THEY DISPEL IGNORANCE OF TIME

I recently watched a hermit thrush, flushed by my clumsy approach. I was walking in a border zone, where redwoods and Douglas-firs intergraded with bays and coast live oaks. Each turn of the winding trail took me to a slope of slightly different grade and orientation, with consequent variation in tree assemblage, groves of redwoods giving way to continuous canopies of thick-leafed live oaks. Of the various habitats in which my local oaks are found, this is the darkest—and therefore the thrushiest.

Strolling unhurriedly along the path, I came upon a large, broken-off redwood limb, lying nearly twenty feet long along the ground, stripped nearly clean of bark by the intervening years since its fall. One part was buried in dry needles; the next segment was richly coated in moss; another patch was lightly sprinkled with green sprouts of poison oak. But the fleeing hermit thrush struck directly for a small area of bare brown wood upon this branch, a backdrop that perfectly matched her middle-brown back, and there she perched, still and silent and alone.

This is a mundane, typical meeting with a winter hermit thrush. At this time of year, they are largely asocial and most commonly encountered singly. And being neither paired nor territorial, they are generally quiet. I very rarely hear hermit thrushes singing in the oak woodland, although a few may do so as spring ripens and they prepare to leave for the tall forests. They can still call in alarm or disquietude: the most common sound you will hear from these birds in winter is the hermit thrush's *chup* note of mild dissatisfaction. (Varied thrushes do sometimes sing in winter, seemingly using song for some communicative function in their nonbreeding flocks, but they are often even quieter than hermit thrushes.)

The winter conduct of birds varies by species. Many flock in large groups, most notably waterfowl and shorebirds, but also many songbirds. In the oak woodlands, several species flock (chickadees, juncos, yellow-rumped warblers), while others are content with solitude (wrens, kinglets). A few woodland birds do maintain year-round pair bonds: jays, titmice, and nuthatches, for instance. What you can say is this: migratory birds will not be in pairs in the winter.

Quietness and a lack of pairs—these are two of the most easily observed traits of the winter thrushes, and two of the most telling. Both indicate the same underlying phenomenon: while "winter" is a convenient human shorthand, biologists would more properly term these thrushes visitors of the "nonbreeding season." According to the calendar, they arrive in fall and leave in spring. What matters is that now is not the time to defend a territory, not the time to find a mate, and not the time to sing.

When I see a thrush in winter, I see that spring's inspiration has receded. But I remember their song, and understand that to recede is not the same as to be extinguished, that there is an ebb and flow in the compulsion to music. To know a songbird in his quiet season is to understand latency and potential, to know a vessel that poured out its contents all summer long and waits now for the spring, when it will once more overflow.

THEY DISPEL THE ILLUSION OF SELF-SUFFICIENT KNOWLEDGE

All the migratory birds of winter demonstrate lessons about space and time, prevent excessive narrowness of my understanding. But note that what this presupposes is accumulated human knowledge —we know where these birds go, and we have broadly understood explanations for their seasonal changes in behavior. Much of this knowledge would be nearly impossible to obtain as a solitary individual. For thousands of years, a debate swung back and forth without resolution: Where do birds go in winter? Do they migrate, hibernate, or perhaps die to be replaced? Our modern understanding of bird migration relies on data collected and shared across wide regions, recoveries of banded birds, and now GPS tracking and other new methods. Part of the wonder of seeing migratory birds is that we now, after millennia of mystery, know where they have travelled from.

Science has a lot to teach about how to see a thrush. I think that human *culture* teaches perhaps even more when it comes to listening to their songs. The words of praise-filled poems and a long heritage of human harmony unavoidably shape how I hear the music of these birds. I listen to both the songs and silences of thrushes, and in the end find that the closer I seek to come to them, the closer I am drawn to the thoughts of other people.

Watching the winter thrushes without knowledge of where they came from, or the reason for their journey, would strip meaning from the meeting, render it opaque and without detail. Listening to these birds unaided by the ears of other listeners would likewise flatten the encounter. Or perhaps prevent it from happening at all—it was books about the birds and trees that first truly drew me to the woods with all my senses open.

And so I see a winter hermit, quietly flicking wings in silence on a tendril of honeysuckle that winds beneath the oaks, and hear that song that others have heard, sounding always in the deep corners of my memory. I watch the varied thrushes rising from the woodland floor to peer down from this modest canopy, and in my mind hear that haunting ringing telling me some strange magic is at work. To know the birds is to have things in your mind that are not apparent to untrained eyes and ears. To know the winter thrushes is to know the deepest secrets of the woods, the journeys from beyond their borders and the songs no longer sung.

CHAPTER 12
A LIGHT IN WINTER

Townsend's Warbler, *Setophaga townsendi*

The winter woods are quiet, cool, and dark. Spring has passed and the trees no longer sing. Fall came and stripped the buckeyes bare, swept south the flycatchers and swallows. The sun does not climb so high as it did before. But in that acute and thin-spread light, a golden spark still kindles.

These woods are quiet, cool, and dark. At this moment, I do not hear a single bird. But I stand arrested as I look into the live oaks and see a moving body of black and brilliant yellow, a silent vision mutely hopping from one branch to another. I watch as the bird comes nearer, see a yellow face bordered by black on the crown above and throat below, and inside the glowing island of his face another deeper sea, a dark triangle on his cheek that recedes and draws my gaze in deeper. And in that geometric black is an eye that glistens from the depth, an eye that sheds a final drop of gold, a tear of spreading feathers that burns like a candle in the dark.

This is the Townsend's warbler, the gem of California's winter. I continue to walk among the oaks when song has ceased, migration has passed, and warm and indoor comforts beckon. I go to look among the leaves for gold, and listen to its silence.

The Quiet Beauty of the Winter Woods

Of all the birds in this book, the Townsend's warbler is the one I associate most strongly with visual, rather than audible, memories.

This is not to say that they never sing or call, but simply that such vocalizations comprise a relatively smaller part of my experience of them as winter birds in California. If you live south of Oregon, the songs of Townsend's warblers are a minor phenomenon, one more wheezy jumble heard for a few weeks amid the April cacophony. And while they do in all seasons utter occasional *chip* calls, they are far less talkative than other winter chippers, like juncos or yellow-rumped warblers, who gather in winter flocks of chatty dispositions. Townsend's warblers in California are essentially neither territorial nor social while I know them, and so taciturnity is their primary vocal characteristic.

To uncover the secrets of the oaks, you have to both listen and look. Winter is increasingly the time for visual observation, when long stretches of near silence give the impression of a deserted canopy—until one of those bustling, mixed-species flocks appears. Sometimes Townsend's warblers join these flocks, and sometimes they forage in solitude. Either way you have to look, searching beyond the squeaking chickadees and the *jit-jit-jit*ering of kinglets to find the woodland wallflower, or combing through the quiet canopy to find the one silent spot of moving gold.

Townsend's warblers are quiet and often decline to announce their presence. In winter, they are thinly spread across California, never qualifying as "abundant" in any of the habitats in which they occur, whether oak woodland, conifer forest, or leafy suburb. So how do you find them? The prescription is simple: spend more time looking at trees.

I think that the frequency of one's encounters with Townsend's warblers is primarily a function of hours spent gazing upon leaves and branches, with the details of habitat selection only a secondary factor. But the oak woodland is still my favorite place to look. The redwood or Douglas-fir forests, for all their grandeur, often keep their birds too high for easy appreciation. Neighborhood birding, for all its convenience, is noisy. But the oaks are human scale without being human dominated: the birds are closer, and when the birds fall silent, so does the world.

So my thoughts are tending on one January day, when rain has coaxed the first milkmaids from the grass and I stroll among the trees. The bark is glowing green with moss and the morning ground gives off a steamy mist. But despite the show of new life emerging from the soil, the birds are yet to awaken from their winter quietude into the noisy jostling of the nesting season, and today there is no song to hear. There are interruptions of sound, as when I pass by a sunny opening where acorn woodpeckers carry on their raucous conversations, or when I step into a bay-shaded ravine where creepers sound sibilantly from the crowding trunks. But overall, the woods are whispering; the beating of their heart has not stopped, but it has slowed, settled into calm quietude as the restlessness of migration and nesting fade into the past.

I stop and gaze at the endless interlocking arches of the canopy. I remove my hat and let the world feel larger, like stepping out of a house to see the sweeping starry sky. I watch the live oaks swaying, their evergreen leaves like sails that always catch the wind, with tiny birds leaping between those swaying masts. I am not searching, not hunting for any particular object, but merely enjoying the breathing of the forest with its quiet, gentle movements. The peacefulness reminds me of watching my wife as she sleeps, in that moment when each first-light exhalation feels like a wordless expression of rightness and belonging.

It is in such moments that the winter warbler quietly appears, a tiny living thing of deep black and boldest yellow that moves among the dark green leaves with an air of resolution. The golden color seems to glow upon his breast like a central moving flame. An eye, alert and searching, looks out from a patch of black surrounded by another gleaming ring of yellow. This bird is not some vaguely idealized presence; he is a spark of real life, with all the individual volition that real life entails. Each wing-assisted leap from one branch to another is jerky and decisive, tending toward the aggressive, while the strong and slightly downcurved beak contributes to a determined impression that dispels any notion of a merely frilly prettiness.

I said that winter is when you need to use your eyes, when hearing becomes less valuable in the search for silent birds. This is true, but incomplete. Listening is just as important now as ever—listening to the silence. The grandeur of cathedrals and the magic of an empty hall of paintings are not incidental to their quietness, but in part derive from it. I think that people love to look at stars because the night is peaceful. The beauty of this warbler stems also from its setting, in time as well as place.

Townsend's warblers bring the northern forests to me, with their taste for trees and the dampening embrace of branches. Townsend's warblers bring the winter too, accompanied inevitably by a relaxation of tension, a greater calm, and an unhurried state of being that nonetheless remains purposeful and self-assured. Winter is not the season of death; it is the season when the core of life is clearest, when the simple spark of awareness and volition is not obscured by sound and fury, nesting and migration.

The winter woods contain a flame of life, air and feathers imbued with flight and motion. Life is here, when many think it sleeping. But quietness is here as well, the deep quietness of winter. This golden bird I watch contains a sea of silent stars.

Seek Out Warblers in the Spring

Townsend's warblers are our most beautiful winter bird. But no bird is simply "a winter bird," materializing in the trees when the weather turns cold. "Winter bird" is simply a shorthand for "doesn't nest here" or "not a summer bird." In truth, birds like Townsend's warblers arrive in fall and linger into spring. Winter birds are migratory birds, and migration is itself a season, a time of movement and excitement, of the sudden arrival of old friends and the passing cameos of transitory visitors. And no birds embody the excitement of migration as fully as the warblers.

What are warblers? One of those old bird books of which I am so fond, Neltje Blanchan's *Bird Neighbors* of 1897, summed up the family as "exceedingly active, graceful, restless feeders among the terminal twigs of trees and shrubbery" who were nonetheless "strangely unknown to all but devoted bird lovers, who seek them out during those months that particularly favor acquaintance." This description captures the two crucial things to know about warblers: that they are birds of the trees, and that you search for them in the migratory seasons, those select months when they are most numerous in the continental states. (It also captures something about people: how do you recognize a devoted bird lover? She is someone who seeks warblers in the spring and autumn woods.)

Consider that first trait, warblers' clear identity as arboreal creatures. This is ecologically very clear: the great majority of warbler species are forest dwellers. And it is central to how we have thought of and categorized these birds. The family Parulidae is more formally known as the *wood* warblers. In earlier times, Townsend's warbler was classified as a member of the genus *Sylvia* (meaning "of the forest") and then *Dendroica* ("tree dwelling"), although modern taxonomy has shuffled them into the less euphonious *Setophaga* ("moth eater").

These old names suggest what matters to us when it comes to warblers. I like to think of Townsend's warblers as members of *Sylvia* or *Dendroica* that were, the sylvans or the tree dwellers. For myself at least, I find such words predispose me to admiration and a vague expectation of magical potential. To step into the forest is to escape from the quotidian contemporaneity of urban existence into timelessness. That is where the warblers live.

Second trait: warblers are migrants. As far as the North American experience goes, nearly all warblers depart in winter, bound largely for Mexico and Central America. As with many tropical migrants, warblers are colorful (most often adorned with splashes of yellow) and eat more animal than plant food. This dietary trend and their particular feeding style of gleaning from the foliage and outer twigs can be read in their physiology: warblers are tiny, lightweight birds with thin beaks that are quite different from the more conical, seed-cracking beaks of goldfinches, our more familiar little yellow birds.

In general, Townsend's warblers fit neatly into this larger family character: they wear yellow, glean insects from the canopy, breed in northern forests, and winter largely to the south of the United States. The thin layer of birds who spend the winter on the Pacific coast of California and Oregon are a wonderful and atypical footnote; the larger share of Townsend's warblers pass their Decembers in a separate population that reaches into southern Texas while primarily extending across a large swathe of Mexico, Honduras, and Guatemala. The existence of this larger non-breeding population in Central America means that Townsend's warblers participate in the great yearly drama of migration, not just in the stealthy arrival of our thinly scattered winter birds, but in the huge passing wave of birds who keep on flying south. Townsend's warblers are the finest birds of California's winter. But they are also migratory birds, birds most abundantly encountered in the fall and in the spring.

In April, I go to the hills near my home. These curves are covered with live oaks, madrones atop the dry ridges, bays on the steeper, north-facing slopes, and valley oaks on the gentle lower reaches. At this time of year, I am surrounded by the songs of local nesters: titmice and juncos, orange-crowned warblers and spotted towhees, warbling vireos and finches. But there are other voices that will soon depart: the mighty tide of warblers is sweeping northward, singing as they go.

There are yellow-rumped warblers, some of whom spent the

winter here, heading for the mountains. There are yellow warblers in search of willows and Wilson's warblers seeking forests. And then there are the three black-throated warblers of the Pacific slope, all discovered by John Kirk Townsend in the Columbia River valley: the black-throated gray, the hermit, and the bird that bears his name, the subject of this chapter. In a month, none of these birds will be found on this hilltop—I would need to travel hundreds of miles to see them, crossing state lines to chase them across the forests of the West. But in spring I simply step out my door and into the oaks, watching and listening as the great wave sweeps them all into my outstretched hands.

Warblers migrate at night. They fly by the stars, enjoy the cool, dark air as their wings pump hour after hour. They don't need the sun to set their course. But they do need it to find food. So, when they look down and see trees as the day breaks, they descend into the oaks and madrones. After hours of continuous flight, they feast with relish upon the burgeoning oak moth caterpillars and all the other fleshy abundance of spring. Insects feed in the bursting catkins on the trees and in the carpet of flowers in their shade; birds pursue them, clinging, hovering, and pouncing. The three black-throated travelers now find themselves among the leaves and warmed by the lengthening days, and so they sing as spring commands them, unfamiliar buzzy songs rising from tree after tree. I attempt to keep up, struggling to distinguish these less familiar songs that fade like flowers, passing northward with the season.

And sometimes I just give up, happily resign myself to indiscriminate enjoyment, give in to sheer profusion. This is the peak of Townsend's numbers, the peak of black-throated grays, the peak of hermit warblers. The thousand dully foraging yellow-rumps of winter transform, now draped in brilliant breeding plumage and joined by new arrivals singing in all their springtime glory. Orange-crowned and Wilson's warblers pour over the hills, filling in the waiting territories like water filling holes as the wave of migration washes over, roaring northward, not exhausting itself for another two thousand miles.

What I witness now in spring is not just a peak in numbers. It is a new side to these birds, the awakening of something dormant. I saw yellow-rumped and Townsend's warblers throughout winter and classed them as songless in my mind. But now I trace unfamiliar music back to birds I thought I knew and realize that there is something more to those figures of my quiet January scenes. In winter, black and gold stood out among the live oaks' dark, unfalling leaves. But now the spark of winter joins in the great blaze of spring's profusion. Now, for this brief moment, this warbler climbs through cascades of flowers; now he sings as shockingly as a painting brought to life.

John Townsend's Treasure

Townsend's warbler, *Setophaga townsendi*, is named after John Kirk Townsend, the first ornithologist to scientifically describe the species after encountering them in the 1830s near the Columbia River. Townsend is unusual in being personally knowable to modern readers in a way that many avian namesakes are not. This is because he wrote the story of his great adventure across the country, publishing it after his return under the catchy title *Narrative of a Journey across the Rocky Mountains, to the Columbia River, and a Visit to the Sandwich Islands, &c., with a Scientific Appendix*. There is no historical ornithologist I feel I know better, and I am glad to remember him in this bird's name. I think that Townsend's era—whatever its flaws, errors, and shortcomings—still has something to teach us about the way we think of nature.

By conventional standards, Townsend is very highly qualified to receive the scientific nod of recognition. He was an excellent and important ornithologist. He introduced over a dozen new species to science and contributed more specimens for Audubon's paintings than anyone else. He was effectively the first ornithologist to find and describe his namesake bird, which is rarely the case. But what is more important is that he was vividly and unquestionably a Devoted Bird Lover in the richest sense of the term. It is easy to forget that the birds we take for granted today—including the quiet black-and-yellow warbler in the woods—were once impossible to encounter without extreme and prolonged efforts, rarely described as vividly as in Townsend's *Narrative*. His story reminds me that such birds are worth the seeking.

John Kirk Townsend was a bird-obsessed young man who left his home of middle-class Philadelphia Quakers in 1834 at the age of twenty-five to join a long expedition across the continent in search of unknown birds. Townsend's tale of his journey remains a very readable story, a tale of dangers and difficulties generally faced with resolve and good humor. On one page Townsend is being chased by an angry grizzly; on another, the company is living off a single dwindling buffalo for a week. The demise of companions and acquaintances from illness, injury, or murder is casually remarked upon at regular intervals. For three years and eight months, Townsend forsook the comforts of home life, left behind his family, church, and everyone he knew, became very sick and nearly died on multiple occasions, lost the majority of his personal belongings, and came back broke. But the reward was immense—he got to see birds.

Not everyone understood this. Townsend's own mother wrote to him a few years into his journey:

> For surely by this time, methinks thou must be tired of dwelling in distant lands, or roaming the Forrests in search of the feathered tribe and the natural curiosities of the Earth, and quite ready to return to civilized life, the comforts of thy own home, and the embrace of those who love thee.

In his *Narrative*, Townsend responded to all those who looked on his enthusiasm for birds with incomprehension:

> They knew but little of the resources of the naturalist; they knew not that the wild forest, the deep glen, and the rugged mountain-top possess charms for him which he would not exchange for gilded palaces; and that to acquaint himself with nature, he gladly escapes from the restraints of civilization, and buries himself from the world which cannot appreciate his enjoyment.

Townsend's era of ornithology was of course far from perfect, and often full of paradox (the frequency of shooting birds in order to admire them is the most obviously jarring note to our well-binoculared age). But the excitement of exploration, the sense that America's wild places held the greatest of all possible adventures and that encountering such birds—*our* birds—was worth an immense amount of danger and material sacrifice—these are feelings that have grown rarer as comforts multiply and wilderness diminishes. But is that loss of excitement inevitable? I still hold onto a belief that incredible things are out there, and that the way to find them is through a lot of walking, among a lot of trees.

Bring a sense of adventure back into your birding. Not through the facsimiles of adventure that anyone with a credit card can obtain—accumulated miles of travel, accumulated birds seen and checked off—but in a sense of the grandeur of your discoveries, however curious they may appear to those at home. Townsend did not journey all those miles so he could brag about his travels. He travelled in order to look at birds, and through them something higher.

In *Some Fruits of Solitude*, William Penn expressed a sentiment that I think Townsend would understand:

> For how could men find the confidence to abuse [the natural world], while they should see the great Creator stare them in the face, in all and every part thereof?

John Kirk Townsend saw the great Creator in the profuse creation of the earth, saw it stare him in the face in every bird he met. This phrasing is Quaker, but the sentiment is nondenominational. The sublimity that Penn identified with the divine can also be simply recognized as that which is beyond our human creation or understanding, that which we could not hope to make and which we therefore should not abuse. And when a person comes to see it in something as simple as a single warbler, striped in black with tears of gold, then their appetite for merely human creations, from cheap trinkets to the most fine-spun of theoretical constructions, will fade as a greater hunger grows.

To me, this bird is not simply "a Townsend's warbler," but John Townsend's warbler, the bird with which he forever wished to be associated. I acknowledge his good judgment each time I speak the name, strive to emulate his passion for life's simple but unending gifts. A bird like this was what he sought, what he travelled all that way to find. I get to see them nearer home—a reason not for indifference, but gratitude. Now I walk among the oaks, in winter and in spring, searching through the leaves to find the object of his journey.

CHAPTER 13
GRAYNESS OVERCOME

Western Bluebird, *Sialia mexicana*

Nuthatches creep down the trunks where acorn woodpeckers cache their winter stores. Towhees and quail scratch on the ground beneath the trees. Chickadees and warblers cling to leaves and probe among the flowers. Creeping, scratching, and clinging are all well enough. But now we come to birds that do what birds do best: fly.

I approach the creek crossing where sycamores clutch the streambed rocks with roots like tentacles, exposed by years of rushing winter water. Valley oaks adorn the surrounding lush green lowlands. On this cool spring day, a still leafless sycamore branch juts from a trunk, and on that thin spar of weathered wood sits a bird, a bird of a richer blue than I have ever seen in the ocean or the sky. He is somewhat plump and unassuming, quietly but alertly watching the ground below with large, thrushy eyes. Tail feathers flick out almost imperceptibly, hinting at a not quite perfect passivity—and then the suggestion is fulfilled with a sudden silent sally toward the grasses below, a passage through the air that sees a brief hover and recalibration before culminating in a pounce into the concealing sea of waving stalks.

The bluebird rises, some indecipherable insect tight-clamped in beak, and flies down the gentle sloping hillside. The ground is covered in vibrant green, with great handfuls of orange poppies strewn across the landscape. Reams of purple vetch crowd the trailside and overhead a red-tailed hawk banks and shows her tail, brick red in the sunlight. And amid this rainbow riot, the bluebird spreads his wings and distills the sky's less concentrated azure. The bluebird spreads his wings and flies, living color moving through the more vivid world of spring.

The Ways They Use the Sky

I think this is the unomittable first fact in the nearly universal human attraction to bluebirds: their color, a rich and vibrant blue that spreads across the head, back, and wings of male birds in a way that's quite unmatched by any equally common bird among the woods. The noisy scrub-jays are dusty and desaturated; forest-dwelling Steller's jays are gorgeous, but comparatively peripheral as birds of the oaks. I have been watching bluebirds for a long time, but I must admit that their blue is still what I think of first when considering the glory of the species, their brilliant blue within a brilliant landscape, their startling vibrance within the vast spread of vibrant colors that surround the springtime trees. Whether bluebirds fly across green and flower-dotted meadows or interrupt the gray skies of winter with their color, people tend to grant them more attention than they do to birds of plainer plumage.

It is also true that this *flying* of bluebirds, that enviable movement right above our heads, attracts our casual attention more than the slower and more camouflaged movements of an earthbound brown chippie or oak-ensconced plain titmouse. I often first spot bluebirds as they string by overhead in family groups, synchronizing with gentle *pew pew* contact calls. At other times, they will launch repeatedly from a tree, launching and retracting with sally after sally like a mosquito-seeking yo-yo. And even when they come down to earth, as they often do in search of insects, they strike through the air, whether from some low perch of observation like a fence line or tree branch, or from a midair hover—it is the rapid movement of flight that attracts our eyes and grabs our attention.

Ecologically, western bluebirds are clearly distinguished in their feeding niche from the scratch foragers, the bark probers, and the foliage gleaners by their frequency of flying. But they resist simple summary, engaging in at least three separate foraging techniques. Least sky-tied, though definitely not earthbound, is their enjoyment of berries, especially the sticky fruit of the mistletoes that clump in the canopies of blue and valley oaks. But most often they eat insects, and most often they catch those insects through a process of spot, sally, seize, rather than through the creeping, clinging, and probing of the bark- and canopy-gleaners. Sometimes this takes the shape of true flycatching, in which flying prey is itself plucked out of the air after a sally from some observation post, but often it involves a descending flight to the ground, in a manner more evocative of a tiny kestrel than of a phoebe.

Bluebirds resist simplistic summary: their ecological niche cannot be encapsulated in a straightforward unitary title along the lines of "the prober of the furrows." Accurately understanding their place among the oaks requires careful investigation, and so all three of their feeding methods merit more attention.

First, the relatively tethered and unflighted one: berry eating. Taxonomically, bluebirds are thrushes, and like varied thrushes, hermit thrushes, and robins, they have a great fondness for berries, far exceeding that of the flycatchers proper. (They also share a reasonable degree of comfort on the ground, though they spend far less time walking about than their other thrush cousins.) While they feed on a number of different berry species, both native and cultivated, bluebirds have a clear favorite in the fruits of mistletoe, the semiparasitic plant that clumps in the canopies of oaks and produces fruit throughout much of the winter and into early spring.

Nuthatch abundance correlates with the abundance of deciduous oaks like blue and valley oaks, in large part because of that bird's affinity for deeply furrowed bark. Bluebirds are found most often among these deciduous oaks too, in large part because these trees are most congenial to the mistletoe species they favor. The most notable of these is *Phoradendron villosum*, or oak mistletoe, which forms shrubby clumps with thick, evergreen leaves in the canopies of trees. Although *Phoradendron* translates literally as "thief of the trees," this and similar mistletoes are not strictly speaking parasites, since they power themselves primarily through their own photosynthesis. Officially, they are semi- or hemiparasites, obtaining primarily water and nutrients from their host trees, while some have argued that they are effectively epiphytes—plants that grow physically upon other plants but have negligible impact upon their health.

In winter, berries from such plants are among bluebirds' most important food sources, and "mistletoe wealth" appears to be one of the primary traits of a "good" bluebird territory. If you see lots of oaks heavily laden with mistletoe, you are probably in a good place to meet bluebirds. It might be more accurate, however, to say that the abundance of oak mistletoe correlates with the presence of bluebirds, rather than the other way around—it is the bluebirds who are the primary dispersers of mistletoe seeds. How do those plants get up into the canopy? Their seeds are not blown up there by the wind. Rather, the berries are consumed by fruit-eating birds (especially bluebirds, but also waxwings, robins, and band-tailed pigeons), enabling the seeds to be carried internally to other trees and then deposited in the birds' droppings on congenial new locations, there to germinate and take hold of their new hosts. As California scrub-jays distribute the oaks, planting acorns in the soil, western bluebirds distribute the mistletoes, planting sticky seeds upon the tree limbs. We have two blue gardeners in the oaks: the loud one sows in soil, while the quiet one is the more aerial bird, and plants her winter stores up among the spreading branches.

The second feeding method of bluebirds is flycatching. This word has two distinct but related meanings in ornithological contexts. On the one hand it is a behavior, encompassing not just any insect-eating, but the specific practice of catching *flying* insects (not gleaning them from surfaces) in *focused* sallies (not gathering up aerial plankton like the constantly flying swallows). But "fly-catcher" can also be used in reference to a distinct family of birds, the Tyrannidae, known as the tyrant or New World flycatchers. This group includes a multitude of species, several of which are found in the California oaks: the ash-throated flycatcher, western kingbird, western flycatcher, western wood-pewee, black phoebe, and Say's phoebe, to name the more common. All of these flycatch as their primary hunting mode, but many other kinds of birds fly-catch at least some of the time, including oak residents like acorn woodpeckers, yellow-rumped warblers, ruby-crowned kinglets, and western bluebirds. Bluebirds flycatch, but are not technically "flycatchers," members of the Tyrannidae.

Flycatching is an important avian foraging method and there are numerous experts of this art among the oaks. When all is told, however, bluebirds have a strong case for being the most noticeable and characteristic flycatching birds in this landscape, despite being only part-time practitioners. Many people notice bluebirds while entirely overlooking the existence of the ash-throats and pewees. Our attention is caught by a bluebird who fractures the immobility of a fence line to dash across the field and pluck a hesitating moth fluttering at head height. We notice a bluebird when she sprouts from a buckeye cavity to seize an unsuspecting bee from its rounds among the flowers, or when she stands atop a nest box, periodically popping up and out or down to grab the mosquitoes unwise enough to appear on her front porch during dining hours. Motion grabs our attention and gorgeous color holds it as the more muted flycatchers rarely do.

The third feeding style of bluebirds is related, but slightly different: the catching on the wing of insects on plants or on the ground. Here, the bluebirds are still flying, but the prey is not. Crickets, grasshoppers, caterpillars, beetles, ants, spiders, snails, and sow bugs may all be quietly crawling along when their blue destiny swoops down from the sky. This technique probably edges out true flycatching as the primary bluebird feeding method, with the relatively grounded prey of grasshoppers and caterpillars comprising the two largest components of their diet. Bluebirds may make their groundward sallies from a natural perch in the trees, a human-provided perch along a fence, or from the self-generated "perch" known as hovering. Such hovering resembles that of kestrels, the little hover-hunting falcons who seek both small rodents and large insects on the ground among the oaks. Both birds love grasshoppers—bluebirds may have more ecological overlap with kestrels than with chickadees and nuthatches.

Bluebirds are not ground birds—they watch from above, even when they approach the soil. They are not bark birds—they do not cling, but dart out through open space. They are not foliage birds, hyperactive and shortsighted—they are patient watchers who use distance as their tool, a span of vacant air they can see through and rapidly traverse. Instead, bluebirds are thrushes, but aerially inclined thrushes who plant the mounds of berries growing above our heads. Bluebirds are flycatchers, the birds among the oaks who we most often notice grabbing a moth out of the sky. And bluebirds are raptors too, sharp-eyed and precise predators who hover where there are no trees to hold them, descending through the air to find the prey more bounded to the earth.

A Mutual Affection

My parents live on the edge of a marsh, a salty expanse extending ultimately to the San Francisco Bay, but bounded on one side by houses, on another by grasslands, and on a third by broadleaf woodlands containing four oak species and a mix of other trees. Their home, in other words, is in perfect bluebird habitat, with oaks and mistletoe nearby, but with plenty of open space for aerial hunting.

And so every year they have bluebirds nesting in their backyard bird houses. Sometimes they are visited by the oak-loving titmice from the nearest stand of trees, and sometimes by the water-loving tree swallows who forage over the marsh. But bluebirds are bird house–loving birds, and come back year after year.

People have a longstanding and widespread affection for bluebirds, who readily reciprocate our attention. When I visit my parents and step out onto their back deck, often enough a bluebird will come flying up to perch on the railing, hopeful that I may be my father, known to them as Bringer of Mealworms. While they continue to make their insect-snapping forays into the air and visit the garden elderberry for fruit, this family of bluebirds looks forward to the dispersal of the mealworms with an appetite unmatched by other birds. As I approach the feeding station, the chestnut-chested male, vibrantly blue above and behind, and his female companion, who echoes the male's colors in a softer and less saturated rendition, sit side by side upon the railing until forced to reluctantly conclude that I have brought them no gifts.

This affection for bluebirds is no quirk of my family alone. It is a phenomenon familiar to countless other wild bird store veterans across the country. Nest boxes are designed with bluebirds in mind, mealworms are carefully raised for bluebird consumption, and various devices are engineered for the express purposes of keeping those bird houses and mealworms out of reach of all the less agreeable creatures of the world. Among the myriad results of capitalistic ingenuity are a surprising array of products designed for the objective of encouraging bluebird reproduction: houses to exclude starlings, houses to discourage house sparrows, entrance portal protectors to protect from jays and crows, poles and baffles to protect from climbing predators, shielded roofs to prevent overheating, caged feeders to exclude larger birds from the precious mealworm allowance—all of these exist because people love their bluebirds.

Even more impressive is the frequent dedication of time:

thousands of people house, monitor, and educate about bluebirds through organizations like the North American Bluebird Society or the California Bluebird Recovery Program. I would not be surprised if the bluebird fanatic website Sialis.org is the most popular taxonomically-derived URL on the internet, luring visitors with crowd-sourced methods for preventing bird house overheating, compilation of at least four authors weighing in on "The Great Entrance-Hole Debate," and every child's favorite online game, "Compare the Nest Boxes." Whether measured in dollars or hours of arcane yet practical research, I can't think of any other avian genus that has elicited this degree of widespread, personal investment in their well-being.

Nor is this attitude some recent phenomenon of softhearted modern consumers. Affection is the default attitude toward bluebirds in all the old bird books I can find. "Who does not welcome the beloved Bluebird and all that his coming implies?" asks Forbush. "Dull indeed would be the man that did not feel the thrill awakened by the first glimpse of [the bluebird's] brilliant color," says Bent. Anyone who appreciates domestic peace and happiness will welcome the "beautiful and singularly loveable" bluebird to their neighborhood, declares Gladden.

The uninitiated might wonder *why* people love bluebirds so much. Beyond their vibrant blue beauty, a common theme brought up by the eastern writers is the bluebird's significance as a herald of spring. (This isn't true of the California birds—we are fortunate enough to have bluebirds all year round.) Many mention their generally sweet and placid temperament. This stands in particularly marked contrast to the jays, our other notably blue birds, who are also notably loud and forceful, especially in contrast to the soft voiced and generally unaggressive bluebirds. And one final factor is their eagerness to accept bird houses, part of their general acceptance of close human association. I think this is important: people love caring for bluebirds because bluebirds are comparatively willing to be cared for. We can take concrete actions to support these birds and see the results in nests and eggs, chicks and fledglings.

It could be argued that humans are more inclined to bluebird-directed altruism because it isn't purely altruism—the presence of bluebirds has generally been beneficial to human endeavors, notably through their consumption of insects. For a significant period of time, such economically quantifiable contributions formed a popular and even dominant lens for evaluating the value of birds. The reason we have studies of songbird stomach contents from the early 1900s is because it was deemed economically important to evaluate which birds consumed agriculturally damaging insects and which birds consumed our crops or beneficial predatory arthropods. With the declining popularity of such "economic ornithology," along with an increasing reluctance to kill hundreds of songbirds for stomach analysis, this type of utilitarian framing has fallen relatively out of favor as a general trend in bird study. But not with bluebirds! The ways in which bluebirds help people continue to be observed, quantified, and publicized, because bluebirds never lack for human champions.

Led by economic ornithology revivalist Julie Jedlicka, researchers have measured the impact of nest box–enhanced bluebird populations in California vineyards and found clear beneficial impacts in reducing agricultural pest species and mosquitoes. While such studies are in some sense a self-interested sector of the ornithological world, they tend to get more popular interest and media attention than "pure science." Self-interest is simply relevance on an easily perceived scale, and I for one am solidly in favor of any kind of avian research that provides the general public with reasons for wanting more birds around.

Dogs and cats used to have jobs, and our residual sense of compatible purpose still contributes to a general attitude of companionship, of being on the same side. The same could be true of birds, and of bluebirds more than most. "Bluebirds are desirable neighbors—

they will eat your mosquitoes and garden pests," you might tell a future bluebird landlord. This is not a distraction from the intrinsic pleasures of their company, but a gateway. Aldo Leopold understood this merger of the practical and the inherently desirable:

> I once counted bird houses visible from the highway and found that only 12 farmsteads out of 100 had bird houses. Did the owners of the other 88 have so much pleasure already that they needed no more?

Bluebirds are the most sought after occupants of bird houses, for reasons of pragmatism (I want them to eat my harmful bugs), ecology (I want to support a potentially declining species), and aesthetic preference (I want to attract the pretty birds). But in the end, I think all bluebird homebuilders revert to an essentially consistent reaction when they see the speckle-chested children emerge and call for feeding from their tireless mother and father: pleasure and joy.

What we are truly seeking is to have that blue around us, the most vivid color of the oaks. Our daily occupations and surroundings are gray enough, and too distant from the world of birds. Our lives are gray enough, and more grounded than we could wish. I think that what we truly seek in the company of bluebirds is a flourishing and wholeness, a clear triumphant siding with these closest fragments of the sky.

CHAPTER 14
THE LIVING WIND

Band-tailed Pigeon, *Patagioenas fasciata*

Pigeons are misunderstood.

Most people who hear the word typically think of feral rock pigeons, the ubiquitous birds of cities around the world. They think—fairly or not—of waddling, slovenly birds that eat crumbs and peck at trash on the sidewalk. They think of birds fluttering ungracefully to the concrete where they trundle among the hurrying city dwellers, who glance at them in brief, disdainful avoidance.

A smaller though not inconsequential number of people have heard of the passenger pigeon, the most famously extinct bird in the history of the continent. By adding that one prefatory word, the conjured image transforms into something entirely different, either a vision of the vast, sky-darkening flocks that once traversed the eastern United States, or simply a pictureless sense of loss, a bare and factual knowledge of a bird that we will never see.

Fewer still, it often seems, are those who know the wild pigeon that still flies among us, the band-tailed pigeon of the West. Unlike the rock pigeons of the cities, these are birds of the trees—birds that roam the woodlands in search of acorns and berries, birds that grip the slender branches of oaks and pass in speeding flocks up and down California and the other western states. Unlike the passenger pigeons, they are animate and alive—they may not darken the sky as those birds once did, but they fly with immeasurably greater force than the passenger pigeons we have today, effigies in museums and old books.

When I hear the word "pigeon," I too think of those two

better-known birds, perhaps respectively the most reviled and the most mourned birds in America. But I think of them as backdrop to the bird more vividly present in my life, as illuminating contrasts that emphasize the qualities that thrill me when I see a flock of band-tails wheeling over the oaks. These are not feral birds, but wild ones, birds that cut the air like ships on trackless seas. These are not the ghosts of birds, but a living wind that roars above the trees, a storm of muscle and feathers.

These Are Not City Pigeons

I descend a richly wooded hillside, long switchbacks passing through the shade of live oaks, black oaks, and madrones. Those last trees, their handsome smooth-barked trunks draped in large, evergreen leaves with gracefully pale undersides, reach up in strong columns to a canopy heavily laden with rough-textured reddish berries. Hearing a weak and curious intermittent call note, I cautiously approach and look up among the leaves in search of the source, a troop of purple finches busily plucking the fruits. Peering up through the crowded branches with my binoculars, however, I now notice a group of much larger birds: a flock of band-tailed pigeons.

They are, like other pigeons, heavy birds, and so they cannot cling to the outer twigs the way the finches do in pursuit of berries. Instead they walk upon the top sides of the branches, inching out until the limbs begin to bend beneath their weight. Then they halt, reaching down and out and back and around, straining their necks to capture every berry within reach from that position of their feet. As they stretch and strain and shift their weight, the branches bend, and sometimes a few wing flaps are necessary to maintain their perch. Five pigeons are in view, all busily at work.

I watch them at length: band-tails are handsome birds, with an air of healthy, clean, wild-roaming vigor that parking-lot pigeons rarely attain. Their form is essentially the same as that of other pigeons and doves (scientifically synonymous words): portly and short legged, small headed and skinny billed. Even while being the most agile and arboreal of the family, band-tails are the largest of my local species, distinctly larger than city pigeons and some three times heavier than mourning doves. Their feathers are an astonishingly smooth dark gray—if you watch for it, you will find that "wild pigeon gray" is one of the most popular modern car colors, a classy yet discreet embodiment of speed. Numerous features clearly distinguish band-tails from other doves and pigeons: the partial white ring on the back of their necks, bright yellow feet, and yellow bills that are tipped in black. Their namesake banded tail is generally inconspicuous, but stands out strikingly if you see them from below with tail spread and sunlight shining from behind. After a brief, hesitant look in my direction, these birds all resume their determined berry plucking.

This is the first trait of forest, as opposed to city, pigeons: they find their food in the trees. Berries are popular, including in California such fruits as those of madrone, manzanita, huckleberry, elderberry, and coffeeberry. But acorns are even more important to band-tails, pursued both in the oaks and on the ground, swallowed whole and digested internally, with the characteristic grinding crop of the family capable of breaking down the outer shells that their weak bills cannot penetrate. Pigeons can't strike with a focused hammer blow, can't break open an acorn the way a jay or acorn woodpecker can. But they can swallow, over and over, and then slowly digest their great feast, like a small snake eating a large gopher, or like me absorbing a whole burrito. The old economic ornithologists kept careful count; examples of crop contents from individual band-tails include 56 whole acorns, 270 whole elderberries, and 622 ponderosa pine seeds. "Pigeon" does not have to mean an urban trash eater: band-tailed pigeons eat the wild fruits of the forest.

Berries and acorns, however, are not consistent, year-round food sources. We've seen a few different responses to this problem: some birds store food, some birds diversify their diet, and some migrate seasonally between northern breeding territories and southern winter territories. Band-tails introduce a new response: nomadism. The flocks gorge themselves on nuts or berries, staying in one area until the trees have nothing left to give, and then they take off, heading to another forest where food will once again be plentiful.

This takes them through many habitats: mixed conifer forests, oak woodlands, and the more widely spaced oak savannas. The wild pigeons do not specialize in open spaces as do the bluebirds, kestrels, and swallows, but they are perhaps most easily seen there, passing overhead without the obstructions of the forest. While band-tails do leave the northern part of their range in winter, their story is not a simple one of north in summer, south in winter. There are band-tailed pigeons in my Bay Area latitudes all year round. But I couldn't say where the band-tail that I saw in August will be in six months.

Band-tails eat until they can eat no more, and then move on. They take off on a journey of hundreds of miles with an easy casualness, trusting in the plenitude of the wide world. And in the tightly contained power of their swiftly flying flocks, there is an immediate drama that even most long-distance migrants lack. They constantly have the air of travelers, of free spirits always ready to open the throttle and leave you far behind.

It is easy to be amazed at the quantifiable feats of the most extreme migratory birds: the bar-tailed godwit flies without food or water for seven days, covering the seven thousand miles from Alaska to New Zealand without a stop! The rufous hummingbird, known to admirers as the "iron-blooded midget," passes across the continent from the chilling Alaskan tundra, buzzes intrepidly over the tossing Caribbean, and cruises at her ease into South America with a bodyweight of one tenth of an ounce! But in everyday life, most migratory birds don't actively proclaim their great feats of flight. They move for a short time and then arrive, their destination reached. Even when their journeys are in progress, many travel at night, and so in their sunlit state appear like any other birds, searching for food in a quite unextraordinary manner. To see a flock of band-tailed pigeons on the move, however, is different. The flight of band-tailed pigeons constantly impresses, proclaims their refutation of the resident and sedentary.

I think there are a few reasons for this perception. Three features of their movements stand out to our eyes: these pigeons fly in flocks, they fly during the day, and they fly with dramatic, distance-devouring speed. For many years, I served as a volunteer counter of the raptor migration over the Golden Gate, where we would regularly see a dozen different hawks and birds of prey. We also saw pigeons. Like the raptors, they were funneled by the contours of the San Francisco Bay. Unlike many of the raptors, their motion across the landscape was forceful and direct. Most of the migrating red-tailed hawks circled and soared at leisure, much like any locally resident red-tail. But band-tailed pigeons would pass by, not as unremarkable individuals, but in flocks of a dozen or a hundred. They would pass by in the day, not seeking the cool air and secrecy of night like most songbirds. And they would pass by with speed and power, eschewing both the leisured enroute feeding of the raptors and the laborious flight of the small-winged songbirds, who need the day to refuel from their labors.

That is how they always move, whether travelling a few hundred yards or a few hundred miles. The great flock lifts off from a tree one morning, to fly and fly and then alight in a new forest and in another world. That wild, unrestricted freedom is the chief trait of these birds as contrasted with the familiar pigeons of the city, which border on the merely feral, given their numerous dependencies on humans. Band-tails are not sidewalk waddlers, objects of our disdain; these are airborne creatures who look down upon us earthbound crawlers from their perches high up in the trees or from their great flights arcing across the ridges, hills, and plains. These are not half-tamed birds inured to our presence as we have become to theirs, but shy and skittish birds who take off with a great furor of wings when a human comes into view.

Still looking up at the canopy of madrones, where the pigeons are clearly aware of my eyes and presence, I lower my binoculars and take another step along the trail. And with that single step I push the pigeons from alertness into alarm. One takes off with a heavy clap of wings, and a split second later the whole flock follows. I had seen five birds, but at least twenty have burst into the air, the loud sound of their pounding wingbeats narrating their departure in a way not heard in the silent flight of smaller birds. The pigeons rise, clattering above the trees, and I watch them disappear, suddenly loud and suddenly vanished, like a caravan of speeding cars on an otherwise empty road, roaring past and disappearing, leaving silence in their wake.

These Are Not Extinct Pigeons

Band-tailed pigeons share their affinity for forest-spanning flight with the passenger pigeon, their closest, but famously extinct relative. What was different about passenger pigeons was the size of their flocks, which numbered in the millions. The written reports of these flocks can seem unbelievable to modern experience, but they are widely attested and substantiated by the quantifiably enormous market that existed for these birds, which were commercially hunted for several decades. One of the more famous descriptions comes from Audubon, who observed a continuous flow of pigeons lasting days on end:

> The air was literally filled with Pigeons; the light of noon-day was obscured as by an eclipse, the dung fell in spots, not unlike melting flakes of snow; and the continued buzz of wings had a tendency to lull my senses to repose. . . . Before sunset I reached Louisville, distant from Hardensburgh fifty-five miles. The Pigeons were still passing in undiminished numbers, and continued to do so for three days in succession. . . . For a week or more the people fed on no other flesh than Pigeons.

Audubon's account corresponds with numerous testimonies from the earliest European explorers through the nineteenth century. One report tells of a breeding colony measured in the early 1800s at over forty miles long, with individual trees laden with over a hundred nests; another colony was recorded in 1871 in Wisconsin covering 850 square miles and containing an estimated 136 million birds. Extrapolating from such reports, modern scientists consider that the passenger pigeon may have once been the most abundant bird in the world, with the great sky-darkening flocks such as those reported by Audubon containing as many as 2 billion birds.

Once upon a time, the passenger pigeon thundered over the beech woods of the East. Now they are gone. What happened? Part of the story involves the loss of much of the original forests of the eastern United States to agriculture and other purposes. Passenger pigeons ate acorns, chestnuts, and beechnuts, and those huge flocks required correspondingly enormous swathes of mast-bearing forest. But a good part of their demise appears to be attributable to the direct, frontal assault of commercial hunting on a vast scale. Such hunting was made especially deadly by the passenger pigeon's habit of gathering in enormous nighttime roosts and breeding colonies, in which hundreds or even thousands of birds could be taken in a single net. The remaining birds were poorly adapted for survival and reproduction outside of those flocks, like city dwellers stuck out in the woods alone and cut off from society's support. Those huge flocks were where they lived and bred, and those huge flocks were easy to destroy: by the late 1800s, only scattered birds remained, and the last passenger pigeon died in a Cincinnati zoo in 1914.

As the population of passenger pigeons diminished, what was lacking was not knowledge of their decline. What was lacking was a scale of values in which the abundant, ongoing existence of pigeons was accorded a high place. In his 1949 elegy for the passenger pigeons, Leopold expresses regret for our collective choosing of "the gadgets of industry" and material comforts over the glory of this bird:

> Men still live who, in their youth, remember pigeons. Trees still live who, in their youth, were shaken by a living wind. But a decade hence only the oldest oaks will remember, and at long last only the hills will know. . . .
>
> Today the oaks still flaunt their burden at the sky, but the feathered lightning is no more. Worm and weevil must now perform slowly and silently the biological task that once drew thunder from the firmament.

Band-tailed pigeons do not sweep through the sky in flocks of millions. This has allowed them to survive while passenger pigeons could not, with their massive breeding colonies unmissable to the hunters and trappers. But the band-tails are nonetheless the greatest flockers of the oak woodlands and savannas. Sparrows, finches, waxwings, or starlings sometimes gather by the hundreds, but none of these can match the force and power of a flock of pigeons at full tilt.

I remember one long walk in the high hills of the Diablo Range in early spring. I was some fifteen miles in and three thousand feet up from the nearest trailhead; the valley oaks of the lowlands, festooned with dangling flowers and bright new leaves, had given way to cooler heights, where widely spaced blue oaks were still largely dormant and patches of snow from the night before lingered on the ground. I had risen from camp at first light and had not seen anyone all day. Continuing my ascent, my path took me past a huge, solitary oak, halfway up a hillside. As I approached, my attention elsewhere, the tree erupted with birds, seeming to launch them skyward from every twisting limb.

Hundreds of band-tails clapped heavily into the air, startled by my appearance in this landscape devoid of walkers. Gray bodies flew haphazardly, a confusion of dark silhouettes against the cloudy sky, impelled by cautious instincts to an alarm that broke the silence and my distracted reveries like a sudden, pelting rain. Birds circled, calmed, and settled gradually back upon the oak's broad limbs, the tree continuing to seethe with the pigeons' nervous caution. The band-tails may not darken the sky as the passenger pigeons once did, but they still can clap like thunder.

And they seem likely to continue to do so. Band-tailed pigeons are not on the verge of extinction: although they have decreased in number over the past several decades, in recent years their population has stabilized and the decline has halted. Hunting in North America is limited and controlled, band-tails are more shy and wary than their extinct cousins, and they do not rely on those huge colonies that left the passenger pigeons vulnerable to tragic tipping points.

Band-tailed pigeons are not on the verge of extinction. I still consider them among the most vivid reminders of that danger, a living evocation of the glories that were lost. Other such glories could be lost again. The world has many marvels to treasure and protect.

But there is also a more pressing danger shown by the case of band-tailed pigeons. Leopold contended that our collective preference for material comfort led to the extinction of the passenger pigeon and left our skies devoid of their awesome rush of wings. His warning still rings true in the case of band-tailed pigeons—

but in a different way. The preference for gadgets and comforts does indeed threaten to rob our lives of wild pigeons, not through imminent ecological collapse, but through the loss of awareness already well in progress. When it comes to this pigeon, the more pressing warning is not to throw away your gadgets, lest the band-tails go extinct, but to throw away your gadgets, lest you miss the ongoing glories of the world. A living wind still flies over the oaks of California; feathered lightning is still striking.

Walking in the woods, I may be wary and alert, or lost in my own private dreaming—in either case, I rarely see the pigeons before they catch sight of me. My usual first awareness of the band-tails' presence comes with the sudden clatter of wings, the explosive liftoff and the crackling of branches. Often I see nothing at all, and am left with no more substantial evidence of their existence than the sound of their departure, a whirring sound that fades like the footfalls of a runner with whom my path did not quite intersect.

But sometimes I catch a glimpse of those heavy bodies, hurtling from the treetops into the unencumbered sky above, the ever-present escape route where they fear no pursuit. A great rush of wings blows over and whistles in the air, rises and falls and fades away like a vanishing dream. The pigeons do not stop for me, and have somewhere else to be. There is no gadget I would take in exchange for that gray blur overhead, for that disappearing clap of wings that sounds like thunder in my mind. The ungraspable essence of the wild is a fleet and fleeting rush of pigeons, a wildness that cannot be contained in cities or in still museum cases.

CHAPTER 15
THE FALCON OF THE OAKS

American Kestrel, *Falco sparverius*

A bright day in May—another one of those glorious spring days when the hills glow with green—and I pass along the not yet exhausted creek, hop from rock to rock at the crossing, and clamber over a tree fallen in the late winter storms, live oak leaves still green. Spring migration is well advanced, with orange-crowned warblers and warbling vireos singing in the canopy, but some birds are still travelling. A cloud of Vaux's swifts swarms overhead, hours of sunlight left before they seek a hollow tree or chimney for the night. And then through this cloud of small, flickering forms comes a larger and more forceful figure.

He overtakes the swifts, powering past with firm strokes until he spreads his wings, pointed and sharply angular, to bank with his tail spread wide. Backlit from above, the finer details of his plumage are obscured, but still the light shines through: a string of white pearls lines the trailing edge of the dark wings and the tail glows with a fiery flicker-red, overlaid by a thick band of black. This is a male American kestrel, the resident falcon of the oaks.

Sometimes kestrels will fly like this, with the classic falcon force and speed. But they are smaller than most falcons—they are in fact the smallest diurnal bird of prey in North America—and will often appear light and buoyant, easily tossed by the wind. At other times they will kite or hover in a quite unfalconlike manner, pinning themselves to the sky as they scan the ground below for grasshoppers or mice. Right now this bird is hunting on the wing among the cloud of swirling swifts, ignoring them in search of smaller prey, perhaps pursuing dragonflies like those I see rising from the creek.

He descends to a treetop across the gorge, wings and tail again flaring briefly as he transitions in a flash from speed to stillness. I raise my binoculars and see the kestrel in profile: wings of nuthatch blue clasped tight around a back of robin-breast rufous, both dotted in uncompromising black. Even from a greater distance, the white cheeks would stand out and identify him, contrastingly sandwiched as they are between two black bars that can be familiarly referred to as the kestrel's mustache and sideburns. But this bird is closer, hiding nothing, and watches back with haughty raptorial superiority, though his fierceness is somewhat diminished by his small dimensions and comic-strip colorfulness.

Such is the kestrel combination: the speed and fierceness of the falcons compressed into a petite and somewhat cartoonishly colorful package. You could think of each of these sides of their character as diminishing the impact of the other—or you can think of both as offering distinct appeal, compartmentalizing ferocity and fun such that both can be appreciated for what they are.

Kestrels are falcons, powerful and predatory. Kestrels are also tiny and colorful, fancy kittens or exotic butterflies in the often brutal world of raptors. This is contrast, not contradiction. Power courses through the oaks, pointed wings scything between trees. Red and blue compress within the smallest of all raptors, a playful portrait perched upon the brown, unyielding limbs.

Kestrels Are Falcons, Fast and Fierce

Most diurnal raptors belong to the family Accipitridae—including red-tailed hawks and Cooper's hawks, kites and harriers, ospreys and eagles. Then there are the falcons, an entirely separate evolutionary lineage that has hit on some of the same traits, such as powerful talons for grasping prey and hooked beaks for tearing flesh.

They are a smaller group, with some forty species worldwide and four in my home in Northern California: the larger peregrine and prairie falcons, and the smaller merlins and American kestrels. All are characterized by a shared form most easily distinguished by their pointed, aerodynamic wings and their consequent fitness for rapid flight (they also tend to have a "mustache" mark).

The classic hunting method of the falcons is high-speed pursuit, although kestrels somewhat qualify this narrative with a few different feeding styles. They don't depend on sudden ambush like the Cooper's hawk and its relatives. They don't specialize in mammal prey with poor eyesight as red-tailed and other broad-winged, soaring hawks often do. Falcons rely on speed, speed that catches songbirds, shorebirds, waterfowl, and any other bird that dares challenge them to a race.

Humans recognize the falcons' speed and power. That's why Falcon is one of the primary protagonists of Native myths of Central California, including stories of the Miwok, Yokuts, and Western Mono. That's why falcons are a favorite of science fiction fans dreaming of interplanetary travel, from Han Solo's ship the *Millennium Falcon* to the SpaceX *Falcon* 9 rockets used to transport astronauts into orbit. People marvel at falcons because they are the fastest creatures on earth, with peregrine falcons measured diving at more than two hundred miles per hour. When people see one of those pointed-wing blurs streak by, it is hard to react otherwise than with a certain degree of gaping wonder.

The evolution of this speed serves a clear purpose: pursuit of prey. The trend of the family toward rapidity in flight suggests their overall inclination toward eating birds, the fast-moving targets who most reward such prowess in life-or-death races. Hawks that hunt short-sighted reptiles and mammals may favor other qualities over speed alone, such as efficiency in soaring or increasingly large size to match heavier mammal prey. Kestrels have not abandoned their family heritage and regularly catch birds, mostly in the wren-to-sparrow weight class, though occasionally punching up into the starling-swallow-meadowlark range. The common falcon names in use until the mid-twentieth century reflect both this familial trend and the prey size preferences within the group: peregrine falcons have been known as duck hawks, merlins as pigeon hawks, and kestrels as sparrow hawks (an identity maintained within their current scientific name, *Falco sparverius*). The current names are traditional too, and not without their counter-justifications—it is true that sparrows are not the *main* food items of kestrels, for instance—but the old appellations concisely convey a significant amount of ecological insight.

Once speed catches a falcon up with her target, then the second falcon feature takes over: fierce, professional efficiency at killing. Most of the true hawks dispatch their prey by clutching their talons repeatedly until resistance ceases. But falcons typically use the propulsion of their flight to power the killing blow, an immediate impact that may stun, disable, or immediately execute their target. For those who survive that initial contact, falcons have a second weapon, a notched bill featuring a tomial "tooth" in the upper mandible that allows for efficient decapitation of small prey and spinal cord cutting of larger animals. A vivid picture of the falcon method is found in Floyd Bralliar's 1922 collection, *Knowing Birds through Stories*, detailing the author's repeated observations of a kestrel preying on young chickens:

> The hawk watches until he feels sure of his prey, then swoops downward straight as an arrow, strikes the bird in the back with his talons, and with his powerful beak tears the top of the head off. The point of the beak is sunk into the base of the skull, and the skull is torn off with a swift forward motion. I succeeded in getting a number of chickens immediately after the hawk struck them, and every one had the whole upper part of the skull torn off, the brain

> exposed, and the medulla mangled with the point of the hawk's beak. After having watched this I felt much less antipathy toward hawks.

Other hawks blindly squeeze over and over, relying on repeated applications of brute force to defeat their frantic prey, like boxers landing punch after punch. But falcons don't fight for sport. They are professional killers: one strike is their minimal norm, and one unerring snip with their beak is their efficient maximum. This particular vein of highly developed efficiency has its harsh and uncompromising side, but also its mercifully brief aspect—the presumed minimization of pain on the prey's part is the reason for Bralliar's decrease in antipathy. Extended suffering is unpleasant to watch, at the least. But a kestrel attack is a performance of clinical precision.

Speed in pursuit and fierce efficiency in execution are the falcons' two main tools, adaptable to many specific prey species. Their specialization in high-speed hunting is not a universally optimal strategy, however, but works best in open areas where birds have room to fly, away from the obstacles and hiding places of the forest. Peregrine falcons often pursue ducks and other waterbirds at the coast. Prairie falcons hunt across plains and deserts. In my area, merlins follow shorebirds migrating south to unobstructed mudflats. Pete Dunne summarizes the family's habitat preference: "wherever the route between points A and B is a straight line."

Which brings us back to the kestrels in the open oak savannas, where they hunt in the fields and grasslands and the wide, open spaces between the scattered trees. Like their other falcon cousins, they disdain the crowded woods and forests of the sharp-shinned, Cooper's, and goshawks. Like the other falcons, they are not confined to the clumsy grounded prey of red-tailed hawks, golden eagles, and white-tailed kites, the main mammal eaters of the oaks, but can twist and turn and accelerate through the air to strike a speeding swallow or clutch a wind-tossed dragonfly. Falcons are the fastest birds and masters of the wind. Kestrels are the falcons in the oaks and rule the tossing billows that blow between the trees.

But Kestrels Are the Smallest Falcons

Raptors are notoriously difficult to identify. Most are hard to approach and seen only at a distance. Many are generally similar in plumage, sharing subtly varying combinations of brownish backs and streaky breasts. But it doesn't take an expert to recognize kestrels.

Kestrel plumage is actually colorful, and different from that of standard-issue hawks. In flight, kestrels are identified by their backs of rusty red with black tiger stripes (this pattern continues on the wings and tails of females). Male birds become even easier to identify, with blue-gray wings contrasting with the rest of their body in a way that is entirely different from any of our other raptors. Even at a distance, perched birds of either sex show their bright white cheeks, unique among small raptors. The classic Bent *Life History* has it right: "The sparrow hawk resembles other birds very little."

In my early days as a hawk migration counter, fresh from my training in flight identification of nineteen different birds of prey, the appearance of even the most common raptors posed difficult-to-resolve questions: Was that *Accipiter* a Cooper's or a sharp-shinned hawk? Was that distant *Buteo* a red-tailed hawk or something rarer? But when a tiny falcon crossed into view, a red-and-blue sickle arcing across the hillside with a blessed lack of ambiguity, I could enjoy the relief of confident self-assertion as I called out the sighting—"One male kestrel!"

As I became more closely acquainted with kestrels, however, my reliance on plumage alone diminished as awareness of shape, flight, and behavior rounded out their portrait. This starts with the most obvious by-the-numbers characteristic of kestrels: smallness.

Generally weighing between three and six ounces, kestrels are more akin in size to a robin or scrub-jay (at the low end) or a California quail (at the high end) than to larger hawks or falcons. Red-tailed hawks weigh as much as ten kestrels, and even peregrine falcons are about five times as large. This results in a windblown buoyancy that nearly all other raptors lack: while kestrels can at times cut through still air with a decidedly falconesque intensity of purpose, they can also struggle into headwinds with a laboriousness reminiscent of smaller songbirds.

Several of the most characteristic traits of kestrels can be seen as stemming from this basic smallness. Their feeding niche, perhaps the most basic element of an ecological portrait, is based around an assortment of prey united by their own even greater diminutiveness. While kestrels do eat birds of sparrow size, their more important food items are small rodents (primarily mice) and large insects (notably grasshoppers, crickets, and dragonflies). "Sparrow hawk" was, for some reason, the name that stayed attached to them longest, but kestrels have also been known as both "mouse hawk" and "grasshopper hawk." And so, while the chick-eating story above was a classically ruthless falcon story, John W. Sugden paints a portrait of a very different tone while observing a kestrel delicately dissecting a grasshopper:

> The grasshopper is held much the same as a child would hold an ice-cream cone. The bird begins by taking several bites of the head. Next the thorax is eaten. The viscera are pulled out and swallowed or occasionally dropped. . . . The first two pairs of legs and the wings are discarded by a flip of the beak. The femora of the third pair of legs are then eaten by taking several bites and the rest of the leg discarded. . . . Small insects are eaten with a similar procedure and rarely by gulping, as the screech owl invariably does.

Kestrels are small: the experience of watching one methodically nibble at a grasshopper reminds the observer of those modest dimensions. Watching bushtits snatch at tiny caterpillars or flickers lapping up ants hardly reads to us as predatory. Kestrels are just one step up in terms of the size of their prey, landing on the ground to snap at spiders with their beaks or twist in midair to clutch at dragonflies. And so, just as vivid cartoon colors diminish overseriousness, I think it is our natural reaction to view a grasshopper eater as less threatening than a bird that slowly tears and dismembers a duck or rabbit. There is, for one thing, far less blood.

Perhaps the most distinctive behavior of kestrels is their frequent hover hunting—this also is enabled by smallness. Many raptors hunt from perches, while many broad-winged hawks of the genus *Buteo*, such as the ubiquitous red-tailed hawks, will also hunt from an efficient soar, slowly circling in the sky without flapping. Kestrels are too small winged to simply launch themselves upon the air to float like a red-tail or turkey vulture. But they are small enough to hover—to actively flap their wings while maintaining a fixed position. This is too energy intensive for most large raptors; imagine the effortfulness of treading water, but in the air. The light weight of kestrels makes this foraging method more practicable, rendering them at times more like bluebirds than like peregrines. As with the bluebirds, kestrels are happy to use perches on trees, wires, poles, and fences, but when no such labor-saving conveniences are available to take them to the middle of a promising hunting field, then they are quite capable of placing themselves exactly where they need to be.

I can remember my early days of watching birds, the days of thrilling ignorance when everything was vivid and novel. I didn't care about rarity in any objective sense; what I cared about was ferocity and power—I was a teenage male, after all—and so the birds I sought were hawks. I would go to the edge of a nearby neighborhood, follow the path that ducked beneath the valley oaks, and climb up the grassy hill to where the raptors hunted. There were red-tails and red-shoulders, harriers and kites—and kestrels. Kestrels powering by overhead, enroute to the grove of oaks across the slope. Kestrels perched on distant treetops, no longer invisible to my newly opened eyes. And kestrels hover-hunting over the fields between the trees, bodies magically in stasis as pointed wings pumped up and down. The falcon came in focus, a silent gaze surveyed the ground below, and then those bottomless black eyes turned toward me, the watcher become watched.

A third consequence—or opportunity—of smallness is the ability to nest in protected tree cavities. On the one hand, this can limit the number of available nest sites in a given area. But, as the success of many cavity-nesting songbirds shows, an enclosed cavity offers superior protection from both weather extremes and predators. Most of the true hawks build stick nests, laboriously constructing a platform that will securely contain their eggs and chicks. But falcons are killers, not builders. Most species either take over nests from hawks or corvids, or use essentially unmodified cliffsides (or concrete ledges, in the case of urban peregrines).

Kestrels mostly don't live by the cliffs and can't fight and displace large nest-building raptors, or even crows. So they look for ready-made cavities, formed either by nature or by large woodpeckers, most often flickers. In these oak-carved hollows, the stories of the bug eaters meet: the ant eaters make the holes and the grasshopper eaters live in them afterward. I am always glad to encounter an occupied cavity, whether the occupant be a titmouse, chickadee, or screech-owl. I savor the trees as the solid bulwarks that they are, living things of both capaciousness and strength. The kestrels too are born inside the trees, while bigger hawks are relegated to the outskirts, farther from the deepest center of the woods. In this as in the rest, smallness brings kestrels closer to the heart of my affections.

Kestrels are falcons, true—swift flyers and fierce killers like the peregrines and merlins. Just as with those larger birds, I thrill to see them cutting through the sky, pointed wings pumping with firm, determined strokes. Were I to see them from the dragonfly's perspective, I would no doubt view their inescapable approach with an appropriate fear at their unwavering glance and firm talons poised to clench tight upon my wings.

Human that I am, however, I cannot help but see in kestrels their very real smallness, their halfway point between the colorful little songbirds that spark feelings of amiable benevolence and the larger clan of killers to which they putatively belong. They are falcons, yes, but falcons that are not dark and dour, but bold in red and blue, with cheeks of bright, clear white beneath those intractable black eyes. They are the little raptors who eat grasshoppers like ice cream cones and thereby shed the raptors' air of ferocity and sternness; they are just slightly larger bluebirds foraging slightly higher over the fields, another splash of vibrance.

And when I watch them hovering before a stand of mighty oaks, I cannot help but remember that they are the falcons who fit inside the trees, the ones who lay their eggs behind a wall of raspy bark. Kestrels are the falcons who emerge like woodpeckers and titmice, taking their first uncertain flight from a doorway cut in wood. Kestrels are the falcons born not of ocean cliffs or desert canyons, but of the trees that hold my sky aloft, the falcons of the oaks.

CHAPTER 16
THE SKY ELF

Violet-green Swallow, *Tachycineta thalassina*

I remember a day when I was very young. The sun was shining, the air was cool but not cold, and a breeze was blowing with an inexpressible freshness. I was outside a childhood home, alone on the lawn, breathing in the scent of spring. And with that air inside me and that tireless wind rushing past my skin, I could not stand still or go inside. I ran across the grass, full of joy to be alive.

I remember another day, when I was young in watching birds, perhaps nineteen or twenty. This was in my youthful raptor days, when I went seeking kestrels and hoping for an eagle. Again it was springtime, again the air was perfect, again I was alone with just the wind and sky. Or so I thought, walking over the green hills amid an open mix of oaks and foothill pines.

But then two violet-green swallows came barreling through the trees, astonishingly vivid figures too close and fast for binoculars. They dipped low over the tall grasses, skimmed that waving surface as they approached me, crossed my path, circled, and returned. Perhaps before they had been hunting, or chasing each other as territorial adults, or chasing each other as children at play—I no longer remember. But I recall them swooping close to investigate this passing stranger, their cheerful chirrups and effortless careening flight. I remember feeling then what I had felt so long before: a simple joy in living, an exuberance and wonder, the delight to be existing in a world of such sounds and smells and colors. Sometimes I think that all I look for in the oaks can be traced back to that day.

Swallows Are the Flying Birds

Most birds have some unique and extravagant story, once you look closely enough: the unmatched force of a kicking spotted towhee, the flicker's tongue that snakes through tunnels in pursuit of ants, the unsuspected constancy with which titmice cling to their partners. What swallows do more magnificently than other creatures is fly.

This is one of the great bird families: the Hirundinidae are found around the globe, with six of the ninety-odd species of the world making a regular appearance in my California hometown. All are characterized by a form that is optimized for flight: long, pointed wings grant them speed and efficiency, legs are reduced to short and nearly superfluous afterthoughts, and broad mouths with small beaks are ideal tools for scooping up tiny airborne insects. A swallow mouth is like a handle-less butterfly net affixed to the nose of a tiny, agile airplane. In my life in the bird-feeding business, we often talk about "ground feeders," or sometimes about "arboreal feeders" if we want to get fancy. Swallows are sky feeders.

In addition to enabling superterrestrial foraging, flying prowess has another life history correlation: it encourages migration. Swallows of temperate latitudes are consistent travelers and tend toward longer journeys than the passerine average. Almost all swallows leave the United States in fall, with several species traveling beyond the typical songbird stopping points in Mexico or Central America to penetrate far into the south. Barn and cliff swallows, for instance, readily migrate between Canada and Argentina: a six-thousand-mile journey broken up into daily flights that can easily exceed two hundred miles. Swallows fly swiftly and tirelessly, finding food as they travel, like Californians boosting their road-trip range with In-N-Out milkshakes. Most birds can't eat as they go, but swallow migration is one continuous drive-through.

Swallows' incredible proficiency at flying is central to the story of how they live. It is also the unmistakable touchstone of almost all of our human encounters with them. The way that swallows slice through the air is not just a fact, but a living phenomenon. No textbook meteorology replaces cold rain lashing at your face, the sun's warmth soaking through your skin, or the wondrous calm of opening your eyes after a night of quiet snowfall. From my corporeal perspective, the flight of swallows is likewise not contained by accounts of feeding habits or migration statistics. In our lives as we live them, the flight of swallows is a rush of speed, a blast of unrestrained existence, a wild blowing away of our tired earthbound lethargy.

Encounters with swallows stand out in my mind like meetings with the most important books, or films, or songs—those moments when another being reaches through your solitude to touch you at your core, crystalizing the surrounding details of time and place into memories suddenly made more vivid. I can think of another spring day when the swallows all flowed northward and I climbed up the tall ridge on the edge of town, where the steepness of the trail discourages most casual walkers and the day's fierce winds deterred even the more vigorous, drowning out my voice and sparking water from my eyes.

But a troop of purple martins, the largest swallows in my area, was more than undeterred—they were exuberant, latching onto the breeze as it crested the ridge, throwing themselves toward me in fearless investigation before banking gently away with effortless control. I felt like I had the privilege of being a falcon's prey, without the discomfort of mortal panic, as I watched their repeated approach. They accelerated inescapably toward me, their streamlined bullet-bodies racing far faster than binoculars could track.

For perhaps five minutes these same half dozen swallows rushed past me, over and over. They took hold of all my attention, became the focal point for my senses bounded by the buffeting gusts. I watched the swallows, and they seemed of one piece with the wind, a wind focused and distilled into bodies of dark, metallic blue. I listened and heard the wind billow past my ears, heard it as the sound of swallows, heard the martins as they rode each cresting wave.

Speed and distant journeys have lost a little luster in an age of cars and planes. But you can still encounter motion as you did before you learned the illusions of machines, as a simple bounded body that walks on clumsy legs. Rediscover the wonder of a life at human speed, climbing to a high point and feeling all your leaden weight. Then gasp and stare at racing swallows, who do not know our limits.

How to Find the Sea-Green Swallow

The core swallow experience described above applies across all species: these are birds that have the wind inside them. But if you want to go beyond that preliminary insight to fully relish that jewel of western swallowdom known as *Tachycineta thalassina*, the necessary starting point is to become a competent recognizer of violet-greens when you see them.

In my home in California, a few simple field marks suffice to attach a label to most swallow encounters: purple martins are uniformly dark all over and rough-winged swallows are dusty brown, while barn and cliff swallows pair dark blue backs with light brown undersides—barns have a forked tail and cliff swallows have a brown rump patch. The more frequent difficulty comes with distinguishing our two "white-bellied swallows," the violet-green and its more widespread relative, the tree swallow. A few pieces of advice can help.

The first and most direct indication is the color of the back: violet-greens are known in Spanish as "la golondrina verdemar" and to science as *Tachycineta thalassina*, the sea-green swallow, for the shockingly vivid emerald green they bear, a luminous and shining encapsulation of exotic oceanic splendor. If light and position allow you to see this green, quite different than the dark metallic blue of the tree swallow, then relish it. Second, if you happen to see a violet-green from above, look for a white rump patch (or, more precisely, two not-quite-joining "saddlebags") above the tail, and—in truly optimal conditions—the subtle violet patch referred to in their name, directly below that white and overlapping the base of the tail.

A third area of relevance comes into play when looking at perched birds: the pattern of their faces. In French, they call our bird "la hirondelle à face blanche," the white-faced swallow—violet-green swallows have large white cheek patches that encompass their eyes. This contrasts with tree swallows, whose dark metallic helmets cover the upper portion of their faces. Tree swallows, like peregrine falcons, follow the Batman school of facial expression: conceal the eyes and cheeks, expose only a pale chin, and maintain an air of constant grim menace.

As with all birds, a large part of the identification process takes place before you ever raise your binoculars, simply by knowing where to look. In the case of swallows, the specificity of their nesting habits and techniques often makes identification possible without the need for difficult observation of plumage details. Swallows returning to a single, mud-based cup under the eave of a building are almost certainly barn swallows. The large colony of mud-built gourds under a highway overpass belongs to cliff swallows. Birds flying into the mouth of a drainage pipe, dry and unused in the California summer, are rough-winged swallows.

And when it comes to the two white-bellied swallows, here too a subtle differentiation suggests an underlying key to the natures of each bird. Both tree and violet-green swallows are cavity nesters: like kestrels and bluebirds, they nest in holes in trees or in human-constructed nest boxes. But tree swallows always nest near water, looking for ponds or reservoirs that form their main hunting grounds. Head into the dry, oak-dotted hills, and the swallows winding their way among the trees will be the violet-greens.

This is the heart of their domain: the oak savanna, where the gnarled and sprawling trees extend their limbs to soak up the vast magnanimous miles of sunlight. The violet-green swallows dart between the oaks, snatching insects rising from the tall grasses, swooping low to fill my eyes at once with green and white and yellow as they fleetingly impress themselves upon a flower-covered canvas. Swallows drop out of a secret hole high up in a two-hundred-year-old oak and pop up effortlessly to weave among the ghost pines. I lay back on a hillside of lupine and poppies and watch the pure bellies of white and glimmers of emerald green flash by in that square of sky above my head like strobe-lit glimmers of a more fantastic world.

Tree swallows are wetland birds, who fly above the water, inaccessible and out of reach. Tree swallows live across the continent; their color spreads across the map. But the sea-green swallows are birds of the land, similarly terrestrial creatures who fly along beside me as I walk among their trees. They are birds of the West, another of those birds that Townsend crossed prairies and mountains to encounter. The violet-green swallows live in the place where I feel most at home.

Emerson noted the "overpowering importance of neighborhood" in determining the quality of our lives. There are only a limited number of people whom we will see at frequent intervals—"these, and these only, shall be your life's companions," he wrote. Choose, as well as you are able, to live near suitable companions who brighten and enrich your life. This truth is not limited to the society of humans—nor to the chickadees he favored! I seek the society of the discontented towhees and the titmice with their oaken bonds, of the quanks patrolling furrows and the greenlets bearing lichens, of the quiet winter thrushes and the silent winter warblers.

And most of all I seek the neighborhood where the sea-green swallows live. When I was young, I walked among the trees alone, until these swallows flew across my path. They danced across the facing hillside as for the first time I poured words into a book. And white faces still watch over me as they ride the wind between the trees.

A More-Than-Human Beauty

> The most beautiful of all the genus hitherto discovered.
>
> —Townsend, *Description of the Birds of the Columbia River Region*, 1836

> The most beautiful Swallow hitherto discovered within the limits of the United States.
>
> —Audubon, *The Birds of America*, 1840

> If we lavished any superlatives on the Tree Swallow, we regret it now. . . . We need all our superlatives for present use. . . . What shall we do for the Violet-green Swallows? Simply this: we will call them children of heaven.
>
> —Dawson, *The Birds of California*, 1923

Violet-green swallows are birds of home, who raise their young within the woods where I would like to do the same. This seems to me a solid starting point for affection: I think it is very natural to be attached to the birds one sees most often. But it would be disingenuous to deny that I place these birds on a rather different plane than other comfortable old companions like titmice and nuthatches, in large part because of the simple fact that the violet-green swallow is the most beautiful bird I know.

Beauty is hard to define. Perhaps one way of getting at the unique appeal of these birds is to compare them just once more to their relatives, the tree swallows. Their primary differences in plumage are in back color—I do find this green more inherently appealing than that blue—and in their differing extent of white upon the face. To focus on white cheeks as a primary trait of extraordinary beauty may seem strange, but Pete Dunne aptly and concisely describes the overall effect of the violet-green visage when he describes their face as "elfin," as possessing a sort of ineffable, wide-eyed, innocent optimism compared to the miserable tree swallows.

Of course, this is not psychologically accurate. But words like "elfin" (and perhaps "grumpy" for the tree swallows) are first of all useful practical shorthands that help to form more broadly memorable and significant characters in our mind. And in this case, the connotations of the word "elfin" fit perfectly with the larger personality of this bird from a natural history perspective. Violet-green swallows do not merely *appear* elfin, but *are* elfin.

There are numerous different versions of elfin character in the realms of folklore and modern fantasy, but the word usually suggests a certain fey and footloose kind of being, nimble and weightless, carefree and alert. Elves are never lugubrious, heavy, lethargic, elephantine. They dance in the wind, sing in voices of playful innocence, and have sharp, unclouded vision. The most prominent elf describer in modern literature is undoubtedly Tolkien, whose elves are not inventions purely out of one man's imagination, but instead draw on an ancient human vision of a not-quite-human race. As Legolas describes a beautiful elf maiden in "The Lay of Nimrodel":

> Her hair was long, her limbs were white,
> And fair she was and free;
> And in the wind she went as light
> As leaf of linden-tree

Fair and free and weightless in the wind—this is an accurate portrait of the violet-green swallows. I watch them rise and fall, living in an element that is nearly imperceptible to my crude senses. I see them pluck invisible sustenance from the invisible currents of the air. And I hear their undamped chatter, a garrulous stream of quick and unreverberant notes that effortlessly come and go, unlike my plodding words.

The beauty of birds is not like the beauty of objects, which often elicits a desire to possess. Nor is the swallows' beauty like the fluffy charm of bushtits and baby quail, which provokes an instinct to protect. For me, their beauty prompts a kind of quiet reverence. "We are not yet half alive to our privileges," said Dawson of life among the violet-greens. The feeling is immediate and direct, eye widening and smile inducing, like stepping into a grove of incredibly tall trees or a meadow filled to bursting with an impossible profusion of flowers.

I think for many people that feeling is imaginable, but rare, requiring some such setting of obvious and overwhelming superabundance. That's not actually the case. Wonder can be contained within a tree, a bird, a flower. A few feathers can contain the green of all the seas and forests; an unclouded face rejoices in the immensity of the sky. And so a swallow can give life to an ancient elfin vision, flying high above us, free from the chains that weigh us down.

A Higher Plane

Part of what draws me to these swallows is their tie to the earth, their tie to the trees and consequently greater interaction and connection with the world I live in: they are born within the oaks and fly through the grasses at my level. That closeness is a wonderful thing. But so is its opposite: their rising up and up to impossible heights. All swallows are creatures of the air, but violet-greens often seem to belong to a higher stratum of existence.

Scientists have noted how violet-greens often feed at more elevated altitudes than other swallows. You can corroborate this tendency with a glance at the family overview page in the Sibley field guides, the most internally consistent collection of North American bird illustrations. Compare the proportions of the different swallows: violet-greens' wings reach out further than their tails to a visibly greater degree than those of their relatives. Long wings are an adaptation for high soaring rather than low-elevation dodging of obstacles. They are red-tails compared to sharp-shinned hawks. Barn and cliff and tree swallows are all nimble, agile beings, but it is rare to watch them continue up and up to utter invisibility.

I love to watch the swallows climb. I love to *hear* them even more, risen beyond the limits of my vision. Swallows are not generally renowned as singing birds, but some species—including violet-greens—perform easily overlooked "dawn songs," often starting hours before daybreak. Most people never come to know the violet-green swallow's music. Most often, we're asleep. If awake, we're usually enclosed behind doors and tight-shut windows. And even if the sounds are not physically cut off from our ears, we might not recognize that continuous chatter as a song, or connect that distant sound floating somewhere up above with the bird we've seen among the trees.

But now you know: the starlight singer over the woods is the loveliest of swallows, long winged and caped in changing green. Their songs sound like a long series of call notes, with each individual note much like their typical chirps, but sewn together into an unending train. The song of the violet-green swallow is not one of the great avian melodies. But it is one of the most *airborne*. What's so special about skylarks, the famous European singers? That they sing from an invisible point in the sky, that one can imagine them as Shelley did, as scorners of the ground, as unbodied joys, and as blithe spirits with whose "clear keen joyance languor cannot be." The violet-green swallow is the skylark of the California oaks.

These songs sound like calls. But they are not. As I lie in bed and hear that keen clear music drifting down, I recognize that these sounds are not serving a merely mundane function of quotidian conversation. They are not some pragmatic communication to a neighbor, as one might be tempted to think during the day's routine foraging. A violet-green swallow awoke today in springtime darkness. And in that black and vernal tension, a force was welling up inside him, a force that told him to rise into the unwarmed night and sing among the stars.

I shed my warm cocoon and step out into the crisp night air. The trees of the day are transformed to quiet shadows, live oaks leaning over me like silent guardians, sheltering the earth from the cold vastness of the sky. My eyes adjust to the darkness and I meet the wary eyes of a fox who thought these woods were hers. I feel the water on the grass and smell the dampness in the soil. When I look up, I see only the slowly turning sea of faint and distant stars—there are no birds in sight. But the world that I can see is not all the world there is. I listen without speaking to a distant, eager chatter, an impatient, unseen heartbeat pulsing in the air. The sea above me simmers as I stand awestruck and amazed.

What birds do I choose to be my life's companions? I choose the undimmed faces who among the oaks have chosen me. I choose the elfin swallows, who show me higher things. And in this swaddling darkness that enwraps our sleeping shadows, I choose the sea-green swallows who summon me by starlight and bid me to awake.

ACKNOWLEDGMENTS

Thank you to the team at Heyday for bringing this book to fruition. Marthine Satris, my editor, constantly pushed me to sharpen both my thinking and my prose. Art director Archie Ferguson embraced and made possible the project of a husband-wife, writer-artist collaboration, to gratifying results. Copyeditor Molly Woodward helped give my sentences elegance and clarity, turning words into a book.

Thank you to my parents, Michael Gedney and Shih-Po Hsu, who brought me first to the woods, then to the library, and eventually to co-ownership of a wild bird feeding shop—the three birthplaces of these essays.

Thank you to the many readers, friends, and customers who shared their stories of the birds, feeding and encouraging my conviction that such stories are worth sharing.

Most thanks of all—several zillions at least—to my ideal illustrator, core of my κοινωνία, and dearest titmouse, Angelina, who brings the birds and so much else to life.

NOTES

Unless otherwise noted, factual information on measurements, range, migration, population status, diet, and contemporary species names draws from the Cornell Laboratory of Ornithology's *Birds of the World*:

S. M. Billerman, B. K. Keeney, P. G. Rodewald, and T. S. Schulenberg, eds., *Birds of the World* (Ithaca, NY: Cornell Laboratory of Ornithology), https://birdsoftheworld.org/.

PREFACE

ix **"Omit the negative propositions"** "Success," in Ralph Waldo Emerson, *Society and Solitude* (New York: E. P. Dutton & Company, 1912), 145.

CHAPTER 1: THE SOUL OF THE OAKS

1 **Blanchan** Neltje Blanchan, *Bird Neighbors* (New York: Doubleday, Page & Company, 1903), 78.

3 **Ten or so unique songs** Song repertoire information and data on countersinging are both found in Keith L. Dixon, "Patterns of Singing in a Population of the Plain Titmouse," *Condor* 71, no. 2 (April 1969): 96, https://doi.org/10.2307/1366070.

3 **"Fighting notes"** Dixon, "Patterns of Singing," 100.

3 **Captain Blood** Rafael Sabatini, *Captain Blood* (Pleasantville, NY: Akadine Press, 1998), 198.

3 **"Song and fighting"** Keith L. Dixon, "Behavior of the Plain Titmouse," *Condor* 51, no. 3 (May 1949): 117, https://doi.org/10.2307/1365105.

5 **Jean Arthur** Dennis Drabelle, "Jean Arthur Revisited," *Washington Post*, October 6, 1991, https://www.washingtonpost.com/archive/lifestyle/style/1991/10/06/jean-arthur-revisited/a2b59f8e-1312-426d-b929-486cc6a80dda/.

7 **The monogamous parid** Jan Ekman, "Ecology of Non-breeding Social Systems of Parus," *Wilson Bulletin* 101, no. 2 (June 1989): 277.

7 **Dickens** Anne Isba, *Dickens's Women: His Great Expectations* (London: Continuum International Publishing, 2011), 41.

7 **John Gibb** John Gibb, "Territory in the Genus Parus," *Ibis* 98,

no. 3 (July 1956), 420–429, https://doi.org/10.1111/j.1474-919X.1956.tb01426.x.

8 **Juniper titmouse** C. Cicero, P. Pyle, and M. A. Patten, "Juniper Titmouse (*Baeolophus ridgwayi*)," *Birds of the World* (March 2020), https://doi.org/10.2173/bow.juntit1.01.

8 **Woodhouse's scrub-jay** Robert L. Curry, A. Townsend Peterson, Tom A. Langen, Peter Pyle, and Michael A. Patten, "Woodhouse's Scrub-Jay (*Aphelocoma woodhouseii*)," *Birds of the World* (March 2020), https://doi.org/10.2173/bow.wooscj2.01.

9 **Shuford** W. David Shuford, *The Marin County Breeding Bird Atlas* (Bolinas, CA: Bushtit Books, 1993), 290.

CHAPTER 2: THE PROBER OF THE FURROWS

15 **Most abundant and successful** Antoine Kremer and Andrew L. Hipp, "Oaks: An Evolutionary Success Story," *New Phytologist* 226 (2020): 987–1011, https://doi.org/10.1111/nph.16274.

15 **More insects and arthropods** The correlation between tree species abundance and number of associated insect species was first presented in the classic paper by Southwood. T. R. E. Southwood, "The Number of Species of Insect Associated with Various Trees," *Journal of Animal Ecology* 30, no. 1 (May 1961): 1–8, https://www.jstor.org/stable/2109. Douglas Tallamy summarizes possible explanations for oaks' extremely high level of arthropod support in Douglas W. Tallamy, *The Nature of Oaks* (Portland, OR: Timber Press, 2021), 40.

15 **Eight-hundred arthropod species** Bruce M. Pavlik, Pamela C. Muick, Sharon Johnson, Marjorie Popper, *Oaks of California* (Los Olivos, CA: Cachuma Press, 1991), 80.

15 **Eighteen oak species** Pavlik et al., *Oaks of California*, 3.

16 **Largest oak** Pavlik et al., *Oaks of California*, 10–11.

19 **Dawson** William Leon Dawson, *The Birds of California* (Los Angeles: South Moulton, 1923), 639.

19 **Pearson** *Birds of America, Part III*, ed. T. Gilbert Pearson (1917; repr., Garden City, NY: Garden City Publishing, 1936), 199.

19 **Forbush** Forbush, *Birds of Massachusetts, Part III* (Norwood, MA: Commonwealth of Massachusetts, 1929), 357.

21 **Covering up insufficiently concealed food** The frequency of such covering appears to be variable, but Petit et al. observed concealment of food in about 50 percent of cases. Daniel R. Petit, Lisa J. Petit, and Kenneth E. Petit, "Winter Caching Ecology of Deciduous

Woodland Birds and Adaptations for Protection of Stored Food," *Condor* 91, no. 4 (November 1989): 768, https://www.jstor.org/stable/1368059.

CHAPTER 3: THE KINGLET AND THE GREENLET

26 **Eye-ringed leaf gleaner** The phrase is borrowed from Rich Stallcup, "The Eye-ringed Leaf Gleaners," *Point Reyes Bird Observatory* (Spring/Summer 1984): 8–9. For Kaufman's consideration of these two species, see Kenn Kaufman, "Identifying Hutton's Vireo," *American Birds* (Fall 1993): 460–62.

27 **Yumph** I am not sure of the exact lineage of this word, but my usage is based on that found in the otherwise rather forgettable William Powell-Myrna Loy comedy, *Double Wedding*.

27 **Commensurate with their smallness** D. L. Swanson, J. L. Ingold, and G. E. Wallace, "Ruby-crowned Kinglet (*Corthylio calendula*)," *Birds of the World* (August 2021), https://doi.org/10.2173/bow.ruckin.01.1.

29 **"Spirit of the live oak"** Clark C. Van Fleet, "A Short Paper on the Hutton Vireo," *Condor* 21, no. 4 (July 1919): 163, https://doi.org/10.2307/1362632. The closeness of Hutton's vireos with live oaks described by Van Fleet reaches its height in California; the association is weaker in other parts of their range. It has been suggested that the discontinuous population found in Mexico and the Southwest states, which appear to be less firmly oak associated, should be considered a separate species from the Pacific populations. See Carla Cicero and Ned K. Johnson, "Genetic Differentiation between Populations of Hutton's Vireo (Aves: Vireonidae) in Disjunct Allopatry," *The Southwestern Naturalist* 37, no. 4 (Dec 1992): 344–48, https://doi.org/10.2307/3671784.

29 **Donald Culross Peattie** Donald Culross Peattie, *A Natural History of Western Trees* (New York: Houghton Mifflin Company, 1950), 450.

30 **Discomfort with the name** "It goes against my principles to name after individuals unless for important scientific service," wrote Hutton in a letter to his uncle. William Rich Hutton, *Glances at California* (San Marino, CA: Huntington Library, 1942), 59.

30 **"Not a bird likely"** Van Fleet, "A Short Paper," 162.

30 **"Distinguished by its demure ways"** Dawson, *Birds of California*, 577.

30 **"I realized that"** Thomas D. Burleigh, "Notes on the Bird Life of

Northwestern Washington," *Auk* 46, no. 4 (October 1929): 55, https://doi.org/10.2307/4076186.

CHAPTER 4: THE RELATABILITY OF THE ABSURD

41 **9 percent of birds** Sahas Barve, Walter D. Koenig, Joseph Haydock, and Eric L. Walters, "Habitat Saturation Results in Joint-Nesting Female Coalitions in a Social Bird," *American Naturalist* 193, no. 6 (June 2019): 2.

42 **Cobreeding** My account of acorn woodpecker breeding systems in California is based on the work of Walter Koenig and colleagues, especially Walter D. Koenig, Ronald L. Mumme, and Dr. Frank A. Pitelka, "The Breeding System of the Acorn Woodpecker in Central Coastal California," *Ethology* 65, issue 4 (1984): 289–308, https://doi.org/10.1111/j.1439-0310.1984.tb00106.x.

44 **Communal food hoarders** Walter Koenig, the leading expert on the cooperative breeding of acorn woodpeckers in California, notes this seemingly central fact, stating that "the dependence of Acorn Woodpeckers on granaries would appear to be critical to their complex social behavior," but heavily qualifies the completeness of this explanation by noting that several Central and South American members of the genus (including some separate populations of acorn woodpeckers) are highly social but do not store food. Overall, however, it remains true that among North American members of the genus, California's acorn woodpeckers demonstrate both the fullest form of cooperative breeding and the greatest degree of communal food hoarding.

For much more detail, see the Cornell *Birds of the World* account by Koenig and his colleagues, which compiles and contextualizes their findings and those of other researchers. W. D. Koenig et al., "Acorn Woodpecker (*Melanerpes formicivorus*)," *Birds of the World* (March 2020), https://doi.org/10.2173/bow.acowoo.01.

44 **Communalism diminishes** "Average breeding-group size ranges from a mean of 4.4 individuals (maximum 15) [in California, while] small groups of 2–3 individuals are the most common breeding units in all populations studied to date." "Although acorn storage occurs throughout the species' range, its extent is variable, being almost universal in California, irregular in southwestern United States, Mexico, and Central America, and apparently only occasional in Colombia." These two observations, along with information on the

uniquely high correlations between acorn woodpecker density, oak density, and oak diversity along the Pacific coast, are elaborated in Koenig et al., "Acorn Woodpecker," *Birds of the World.*

45 **At least two different oak species** Walter D. Koenig and Joseph Haydock, "Oaks, Acorns, and the Geographical Ecology of Acorn Woodpeckers," *Journal of Biogeography* 26, no. 1 (January 1999): 159–165, https://www.jstor.org/stable/2656045.

45 **Fifty thousand holes** Dawson, *Birds of California*, 1028.

45 **House in the Sierras** William Emerson Ritter, *The California Woodpecker and I: A Study in Comparative Zoology* (Berkeley, CA: University of California Press, 1938), 124.

45 **Between 50 percent and 60 percent** Koenig et al., "Acorn Woodpecker," *Birds of the World.*

46 **A mix of species** Koenig and Haydock, "Oaks, Acorns, and the Geographical Ecology of Acorn Woodpeckers," 165.

46 **I've written elsewhere** Jack Gedney, *The Private Lives of Public Birds: Learning to Listen to the Birds Where We Live* (Berkeley, CA: Heyday, 2022), 13.

47 **"One of the most pathetic"** Dawson, *Birds of California*, 1029.

CHAPTER 5: DISCONTENTED SHADOWS

53 **John Davis** John Davis, "Comparative Foraging Behavior of the Spotted and Brown Towhees," *Auk* 74, no. 2 (April 1957): 142.

53 **"He rustles the dry leaves"** Edward Howe Forbush, *Birds of Massachusetts, Part III* (Norwood, MA: Commonwealth of Massachusetts, 1929), 108.

53 **"I often expected"** Henry David Thoreau, *Thoreau on Birds* (1910; repr., Boston: Beacon Press, 1993), 353.

54 **Proportions of their leg bones** John Davis, "Comparative Foraging Behavior of the Spotted and Brown Towhees," *Auk* 74, no. 2 (April 1957): 163.

55 **"Chuwheeo"** This name from Virginia is recorded in Bayard H. Christy, "Topsell's 'Fowles of Heauen,'" *Auk* 50, no. 3 (July 1933): 278.

55 **"Nasty, whiny, rising-and-falling"** Pete Dunne, *Pete Dunne's Essential Field Guide Companion* (New York: Houghton Mifflin Harcourt, 2006).

56 **"Two globes of crimson fire"** William Blake, "The Marriage of Heaven and Hell," in *The Poetry and Prose of William Blake* (Garden City, NY: Doubleday & Company, 1965), 40.

56 **"Mighty Devil folded in black clouds"** Blake, *Poetry and Prose of William Blake*, 35–37.

57 **"General dissatisfaction"** Joan Beltz Roberts, "Vocalizations of the Rufous-sided Towhee, *Pipilo erythrophthalmus oregonus*," *Condor* 71, no. 3 (July 1969): 261.

59 **Trilling more than six hundred times per hour** John Davis, "Singing Behavior and the Gonad Cycle of the Rufous-Sided Towhee" *Condor* 60, no. 5 (September–October 1958): 320. My conservative estimate of 5,000 songs for the day extrapolates from Davis's measurement of 3,390 songs heard in eight hours, a figure that excluded six daylight hours, including the first two hours of the morning.

59 **Decline in male song** Davis, "Singing Behavior," 322.

CHAPTER 6: THE GENTLENESS THAT BINDS

67 **The male will lose measurable weight** A. Starker Leopold, *The California Quail* (University of California Press, Berkeley, CA: 1977), 69, 89.

67 ***Put-put* calls** For a more complete summary of vocalizations, see Leopold, *California Quail*, 70–71.

67 **Native names** Leopold, *California Quail*, 227.

68 **Less vulnerable to jays and crows** See, for instance, the work of Michael Patton and Douglas Bolger analyzing comparative predation risks to ground-nesting and shrub-nesting birds in California coastal scrub. In their study area, none of the predation events on ground-nesting birds were attributed to avian predators (snakes were the most common nest predator). Michael A. Patten and Douglas T. Bolger, "Variation in top-down control of avian reproductive success across a fragmentation gradient," *Oikos* 101, no. 3 (June 2003): 485.

68 **Most notably coyotes** Kelly J. Iknayan, Megan M. Wheeler, Samuel M. Safran, Jonathan S. Young, Erica N. Spotswood, "What makes urban parks good for California quail? Evaluating park suitability, species persistence, and the potential for reintroduction into a large urban national park," *Journal of Applied Ecology* 59, no. 1 (January 2022): 199–209.

69 **Dwight Huntington** Quoted in Arthur Cleveland Bent, *Life Histories of North American Gallinaceous Birds* (1932; repr., New York: Dover, 1963), 68.

69 **Noticeably less numerous** This section detailing the historical abundance, subsequent decline, and early attempts at recovery is drawn primarily from Leopold, *California Quail*, 25–44.

70 **One hundred thousand each year** Bent, *Life Histories of North American Gallinaceous Birds*, 68.

71 **Han-shan** The line is not a direct quote from Han-shan (Cold Mountain), but from the inscription on a famous portrait of Han-shan and Shih-te by Luo Ping, much reproduced via the ink rubbing of Tang Renzhai. For a full translation, see Kim Karlsson, Alfreda Murck, and Michele Matteini, *Eccentric Visions: The Worlds of Luo Ping* (Museum Reitberg Zurich: 2009), 275.

71 **Montaigne** Michel de Montaigne, "De l'institution des enfants," in *Les Essais* (1595; repr., Paris: Le Livre de Poche, 2001), 248.

CHAPTER 7: A BIRD AT OUR LEVEL

76 **One Chippewa name** W. W. Cooke, "Bird Nomenclature of the Chippewa Indians," *Auk* 1, no. 3 (July 1884): 244, https://doi .org/10.2307/4066840.

77 **Aristotle** Aristotle, *History of Animals*, trans. D'Arcy Wentworth Thompson (Oxford: Clarendon Press, 1910), Bekker number 9.11,615a15–20.

77 **Japanese folktale tradition** Yanagita Kunio, "Japanese Folk Tales," trans. Fanny Hagin Mayer, *Folklore Studies* 11, no. 1 (1952), 5–6, http://www.jstor.org/stable/1177324.

78 **Within ten feet** Judith L. Wagner, "Seasonal Change in Guild Structure: Oak Woodland Insectivorous Birds," *Ecology* 62, no. 4 (August 1981), 974–7, http://www.jstor.org/stable/1936996.

79 **Jenny Wren Spotless Food Stores** Jenny Wren Spotless Food Stores, "Why Are They Called Jenny Wren Stores?" advertisement, *Oakland Tribune*, January 28, 1927, 17.

79 **"Myth is invention about truth"** Humphrey Carpenter, *Tolkien: A Biography* (Boston: Houghton Mifflin Company, 1977), 147.

81 **This bird's namesake** As of this writing in early 2024, the American Ornithological Society has announced their intention to change the common names of all birds named after people in the coming years. Bewick will continue to be recognized in the scientific name, *Thryomanes bewickii*.

81 **Audubon** Many paint Audubon as competitive and egoistic. But in the extensive comments he made regarding Bewick, a sincere

admiration seems to shine through. "So I shall see and talk with the wonderful man. I call him wonderful because I am sincerely of the opinion that his work on wood is superior to anything ever attempted in ornithology," wrote Audubon on the eve of their meeting. John James Audubon, *Audubon and His Journals Vol.* 1 (London: John C. Nimmo, 1898), 231.

83 **Bewick's wrens' claim to fame** This distinction is shared in the East with the Carolina wren, another nester in strange places.

83 **In the record books** The collected list of nest sites is compiled from Dawson, *Birds of California*, 673; Arthur Cleveland Bent, *Life Histories of North American Nuthatches, Wrens, Thrashers, and Their Allies* (1948; repr., New York: Dover, 1964), 177–178; and Florence A. Merriam, *A-Birding on a Bronco* (Boston: Houghton Mifflin, 1896), 95.

83 **Motorcycle helmet** Robin McKay in discussion with the author, May 2023.

83 **12 percent of nests** P. C. Bibbee, "The Bewick's Wren (Thryomanes bewickii (Audubon))" (Ph.D. thesis, Cornell University, 1947) quoted in E. D. Kennedy and D. W. White, "Bewick's Wren (*Thryomanes bewickii*)," *Birds of the World* (March 2020), https://doi.org/10.2173/bow.bewwre.01.

83 **Abandoned automobiles** R. W. Campbell et al., *The Birds of British Columbia Volume 3: Passerines—Flycatchers through Vireos* (Vancouver, BC: University of British Columbia Press, 1997) quoted in E. D. Kennedy and D. W. White, "Bewick's Wren."

CHAPTER 8: THE LITTLE TAILOR

88 **Smallness of the twigs** Wagner, "Seasonal Change in Guild Structure: Oak Woodland Insectivorous Birds," 976.

88 **Lowly midgets** Bent, *Life Histories of North American Jays, Crows, and Titmice*, 449.

88 **Dun-colored atoms** Dawson, *Birds of California*, 631.

89 **67 percent of the time** Paul E. Hertz, J. V. Remsen, Jr., and Stacey I. Zones, "Ecological Complementarity of Three Sympatric Parids in a California Oak Woodland," *Condor* 78, no. 3 (Autumn 1976): 310, http://www.jstor.org/stable/1367689.

89 **"Cute objects"** Sianne Ngai, "The Cuteness of the Avant Garde," *Critical Inquiry* 31, no. 4 (Summer 2005): 815–6, http://www.jstor.org/stable/10.1086/444516.

89 **Plump fluffy tot** Merriam, *A-Birding on a Bronco*, 103.

90 **"Clumped" roosting** Susan M. Smith, "Roosting Aggregations of Bushtits in Response to Cold Temperatures," *Condor* 74, no. 4 (Winter 1972): 478, http://www.jstor.org/stable/1365903.

90 **Perfectly warm weather** S. A. Sloane, "Bushtit (*Psaltriparus minimus*)," *Birds of the World* (March 2020), https://doi.org/10.2173/bow.bushti.01.

90 **Couples snuggle together** Alice Baldwin Addicott, "Behavior of the Bush-tit in the Breeding Season," Condor 40, no. 2 (March–April 1938): 59, https://doi.org/10.2307/1363843.

93 **Unrelated individuals** Sarah A. Sloane, "Incidence and Origins of Supernumeraries at Bushtit (*Psaltriparus minimus*) Nests," *Auk* 113, no. 4 (October 1996): 757–770, https://doi.org/10.2307/4088855.

94 **"The most beautiful bird houses"** Dawson, *Birds of California*, 635.

95 **As many as fourteen birds** S. A. Sloane, "Bushtit (*Psaltriparus minimus*)," *Birds of the World.*

95 **Over three hundred separate feathers** Merriam, *A-Birding on a Bronco*, 107.

95 **Full Arizona sunlight** S. A. Sloane, "Bushtit (*Psaltriparus minimus*)," *Birds of the World.*

CHAPTER 9: A FIRE RISES

102 **Numerous component adaptations** For a clear and well-illustrated overview of woodpecker adaptations, see David Allen Sibley, *What It's Like to Be a Bird* (New York: Penguin Random House, 2020), 87.

103 **Flickers are the most important** For a representative example of flickers' importance, see the work of Martin et al., in which ten different animal species were found using flicker cavities in a single forest community: deer mouse, short-tailed weasel, red squirrel, northern flying squirrel, starlings, tree swallow, mountain bluebird, saw-whet owl, kestrel, and bufflehead. Kathy Martin, Kathryn E. H. Aitken, and Karen L. Wiebe, "Nest Sites and Nest Webs for Cavity-Nesting Communities in Interior British Columbia, Canada: Nest Characteristics and Niche Partitioning," *Condor* 106, no. 1 (February 2004): 5–19, https://doi.org/10.1093/condor/106.1.5.

104 **50 percent of their diet** F. E. L. Beal, *Food of the Woodpeckers of the United States* (Washington, DC: United States Department of Agriculture, 1911), 59–60.

104 **Five thousand individuals** Beal, *Food of the Woodpeckers*, 54. The five-thousand-ant stomach belonged to a yellow-shafted flicker, whose food habits were overall very similar to the western red-shafted subspecies.

104 **20 quadrillion ants** Patrick Schultheiss, Sabine S. Nooten, Runxi Wang, Mark K. L. Wong, François Brassard, Benoit Guénard, "The abundance, biomass, and distribution of ants on Earth," *PNAS* 119, no. 40 (2022): 57–65, https://doi.org/10.1073/pnas.220155011.

105 **Two and a half inches** Dawson, *Birds of California*, 1046. Dawson, writing in the 1920s, notes that you can actually achieve three inches if extending the tongue by hand. Ornithology was different then.

106 **Willard Ayres Eliot** Willard Ayres Eliot, *Birds of the Pacific Coast* (1923; repr., London: Forgotten Books, 2015), 118–19.

107 **Forbush's classic** Edward Howe Forbush, *Birds of Massachusetts, Part II* (Norwood, MA: Commonwealth of Massachusetts, 1927), 293.

107 **William Leon Dawson** Dawson, *Birds of California*, 1041.

107 **Over 160 different names** 160 names are included in James Kedzie Sayre, *North American Bird Folknames and Names* (Foster City, CA: Bottlebrush Press, 1996), 90. I have found two additional names (including the notable "dishwasher") recorded in Frank L. Burns, "Additional Vernacular Names of the Flicker (*Colaptes auratus*)," *Wilson Bulletin* 22, no. 1 (March 1910): 55, https://www.jstor.org/stable/4154286.

108 **Frank Burns** Frank L. Burns, "A Monograph of the Flicker (*Colaptes auratus*)," *Wilson Bulletin* 7, no. 2 (April 1900): 11.

108 **Miwok stories** C. Hart Merriam, *The Dawn of the World* (Cleveland: Arthur H. Clark Company, 1910), 49 and 111.

108 **Ceremonial headdresses** Edward Winslow Gifford, "Miwok Cults," *University of California Publications in American Archaeology and Ethnology* 18, No. 3 (1926): 406.

108 **Not a rare and merely local phenomenon** Gordon W. Hewes, "Californian Flicker-Quill Headbands in the Light of an Ancient Colorado Cave Specimen," *American Antiquity* 18, no. 2 (October 1952): 147–54.

CHAPTER 10: THE ANTIDOTE OF FEAR

111 **Combinatory languages** Simon Harrap and David Quinn, *Chickadees, Tits, Nuthatches, and Treecreepers* (Princeton, NJ: Princeton

University Press, 1996), 267. The language of black-capped chickadees is the most well studied, but similar patterns of meaning have been documented in Mexican chickadees and seem safe to presume, at least on some level, among other close relatives with similar vocalizations.

111 ***Chick-dee-dee* means *great gray owl*** Christopher N. Templeton, Erick Greene, and Kate Davis, "Allometry of Alarm Calls: Black-Capped Chickadees Encode Information About Predator Size," *Science* 308, no. 5730 (June 2005): 1934–37, https://doi.org/10.1126/science.1108841.

114 **More time in the foliage** Paul E. Hertz, Remsen, and Zones, "Ecological Complementarity of Three Sympatric Parids," 310.

115 **Flocking tendencies increase** The two main theoretical rationales for flocking are improved foraging efficiency and increased predator avoidance. As northern forest birds who are very small, chickadees derive a significant benefit in both of these mutually compatible models and are the central members of mixed-species flocks in northern hemisphere forests. For more on both this role in general and the influence of temperature on the flocking impulse, see Bert C. Klein, "Weather-Dependent Mixed-Species Flocking During the Winter," *Auk* 105, no. 3 (July 1988): 583–84, https://doi.org/10.1093/auk/105.3.583.

116 **"All that is evergreen"** Thoreau, *The Journal*, November 7, 1858 in *Thoreau on Birds*, 421–22.

116 **Ralph Waldo Emerson** The poem is "The Titmouse"; the subject is the black-capped chickadee. At that time, bird names were less standardized and "titmouse" was a perfectly acceptable designation. Ralph Waldo Emerson, *Emerson's Prose and Poetry* (New York: W. W. Norton & Company, 2001), 472.

119 **"We interfere with the optimism"** Emerson, "Spiritual Laws," *Emerson's Prose and Poetry*, 151.

CHAPTER 11: THE WINTER THRUSHES

124 **Florence Merriam Bailey** Florence A. Merriam, *Birds Through an Opera-Glass* (Cambridge, MA: Houghton, Mifflin & Company, 1889), 197.

125 **The most praised song** Emily Doolittle identified ninety-seven published poems wholly or partially about hermit thrush songs between 1865 and 1940. Emily Doolittle, "'Scarce Inferior to the

Nightingale': Hermit Thrush Song as a Symbol of Cultural Identity in Anglophone North America," *Ecomusicology Review* 8 (2021): https://ecomusicology.info/doolittle21/.

125 **"This bird never fails"** Thoreau, *The Journal*, July 5, 1852, in *Thoreau on Birds*, 429. Although the full passage refers to the "wood thrush," Thoreau did not distinguish between that species and the hermit thrush. This passage seems to me likely to refer to the latter bird.

126 **"Changes all hours"** Thoreau, *The Journal*, June 22, 1853, in *Thoreau on Birds*, 431.

126 **"Seems to be the voice"** John Burroughs, *Wake-Robin* (Edinburgh: David Douglas, 1884), 70.

126 **"He who has not"** Dawson, *Birds of California*, 748–49.

126 **"One forgets all trivial things"** William Leon Dawson, *The Birds of Washington Vol.* 1 (Seattle, WA: Occidental Publishing, 1909), 241.

128 **Pieplow** Nathan Pieplow, *Peterson Field Guide to Bird Sounds of Western North America* (New York: Houghton Mifflin Harcourt, 2019), 371.

130 **Varied thrushes continue to eat** F. E. L. Beal, *Food of the Robins and Bluebirds of the United States* (Washington, DC: United States Department of Agriculture, 1915), 17.

130 **Most often on the ground** When not feeding on berries, 87 percent of observed varied thrushes fed on the ground in a breeding ground study. Maurie J. Beck and T. Luke George, "Song Post and Foraging Site Characteristics of Breeding Varied Thrushes in Northwestern California," *Condor* 102, no. 1 (February 2000): 98, https://www.jstor.org/stable/1370410.

CHAPTER 12: A LIGHT IN WINTER

141 **"Exceedingly active"** Blanchan, *Bird Neighbors*, 10.

143 **John Kirk Townsend** As in the case of the other eponymously named birds in this book, the common name of Townsend's warbler is slated to be replaced at some point in the coming years. The scientific name *Setophaga townsendi*, however, will continue to recognize Townsend.

144 **Flaws, errors, and shortcomings** The flaws of historical ornithologists have been widely discussed of late. The American Ornithological Society's 2023 decision to rename all North American birds named after people came after years of campaigning by activists who

emphasized the dissonance between modern views and those of the historical honorees. Personally, I am not convinced that changing bird names is in itself a substantive act of social justice, rather than a prime example of one of the "affectively charged flashpoints" that distract from actual work to reduce present-day material inequities, skillfully described in Liam Kofi Bright, "White Psychodrama," *Journal of Political Philosophy* vol. 312 (June 2023): 192–221.

And so, while I support the project of replacing eponymous bird names with descriptive names—I think the change will almost certainly average out as beneficial to the goal of having more people learn the names of birds—I do not share the original campaigners' aversion to discussing the positive aspects of, for example, John Kirk Townsend's life and legacy.

144 **The first ornithologist to find** Townsend's warbler was officially named by Thomas Nuttall, the senior naturalist on the expedition. But given that Townsend was charged with all ornithological responsibilities and named all other birds reported from the journey, it seems likely that Nuttall's role was merely to satisfy the conventional prohibition against naming a bird after oneself.

145 **"Surely by this time"** Priscilla Townsend to her son, March 3, 1836. Cited in Barbara and Richard Mearns, *John Kirk Townsend: Collector of Audubon's Western Birds and Mammals* (Dumfries, Scotland: Barbara and Richard Mearns, 2007).

146 **"They knew but little"** John Kirk Townsend, *Narrative of a Journey across the Rocky Mountains, to the Columbia River, and a Visit to the Sandwich Islands, &c., with a Scientific Appendix* (Corvallis, OR: Oregon State University Press, 1999), 157.

147 **"For how could men"** William Penn, *Some Fruits of Solitude* (Philadelphia: Benjamin Johnson, 1792), 5.

CHAPTER 13: GRAYNESS OVERCOME

152 **Effectively epiphytes** Walter D. Koenig, Johannes M. H. Knops, William J. Carmen, Mario B. Pesendorfer, and Janis L. Dickinson, "Effects of mistletoe (*Phoradendron villosum*) on California oaks," *Biology Letters* 14, no. 6 (June 2018): https://doi.org/10.1098/rsbl.2018.0240.

152 **Mistletoe wealth** Janis L. Dickinson and Andrew McGowan, "Winter resource wealth drives delayed dispersal and family-group living in western bluebirds," *Proceedings of the Royal Society* 272, no.

1579 (September 2005): 2423–28, https://doi.org/10.1098/rspb.2005.3269.

153 **Grasshoppers and caterpillars** Grasshoppers and caterpillars were found to be the two most popular food items in Beal's 1915 analysis of mostly Californian birds, each contributing slightly over 20 percent of the annual diet. Beal, *Food of the Robins and Bluebirds*, 27.

156 **Forbush** Forbush, *Birds of Massachusetts, Part* III, 419.

156 **Bent** Arthur Cleveland Bent, *Life Histories of North American Thrushes, Kinglets, and Their Allies* (1949; repr., New York: Dover, 1964), 234.

156 **Gladden** George Gladden, "Bluebird," in *Birds of America, Part III*, ed. T. Gilbert Pearson (1917; repr., Garden City, NY: Garden City Publishing, 1936), 242–43.

157 **Agricultural pest species** Julie A. Jedlicka1, Deborah K. Letourneau, and Tara M. Cornelisse, "Establishing songbird nest boxes increased avian insectivores and reduced herbivorous arthropods in a Californian vineyard, USA," *Conservation Evidence* 11 (September 2014), 34–38, https://conservationevidencejournal.com/reference/pdf/5485.

157 **Mosquitoes** Julie A. Jedlicka, Anh-Thu E. Vo, and Rodrigo P. P. Almeida, "Molecular scatology and high-throughput sequencing reveal predominately herbivorous insects in the diets of adult and nestling Western Bluebirds (*Sialia mexicana*) in California vineyards," *Auk* 134, no. 1 (January 2017): 116–27, https://doi.org/10.1642/AUK-16-103.1.

158 **Leopold** Aldo Leopold, "Bird Houses" in *For the Health of the Land* (Washington DC: Island Press, 1999), 130.

CHAPTER 14: THE LIVING WIND

163 **Crop contents** D. M. Keppie and C. E. Braun, "Band-tailed Pigeon (*Patagioenas fasciata*)," *Birds of the World* (March 2020), https://doi.org/10.2173/bow.batpig1.01.

164 **"Iron-blooded midget"** Dawson, *Birds of California*, 931.

167 **"The air was literally filled"** John James Audubon, *The Birds of America Vol.* 5 (1840; repr., New York: Dover, 1967), 27.

167 **Forty miles long** Edward Howe Forbush, "Passenger Pigeon," in *Birds of America, Part II*, ed. T. Gilbert Pearson (1917; repr., Garden City, NY: Garden City Publishing, 1936), 40.

167 **850 square miles** David Allen Sibley, *The Sibley Guide to Bird Life & Behavior* (New York: Knopf, 2001), 321.

167 **2 billion birds** Sibley, *Sibley Guide to Bird Life & Behavior*, 324.

168 **"Men still live"** "On a Monument to the Passenger Pigeon," in Aldo Leopold, *A Sand County Almanac, and Sketches Here and There* (1949; repr., New York: Oxford University Press, 1989), 109–11.

169 **They have decreased** M. E. Seamans, "Band-tailed Pigeon Population Status 2022," *US Department of the Interior, Fish and Wildlife Service*, August 2022, https://www.fws.gov/library/collections/band-tailed-pigeon-population-status-reports.

CHAPTER 15: THE FALCON OF THE OAKS

175 **Native myths** Prairie Falcon is a central figure in the stories of the Miwok, Yokuts, and Western Mono. For the Miwok, see Edward Winslow Gifford, *Miwok Myths* (Berkeley, CA: University of California Press, 1917). For the Yokuts, see A. L. Kroeber, *Indian Myths of South Central California* (Berkeley, CA: University Press, 1907). For the Western Mono, see Edward Winslow Gifford, "Western Mono Myths," *Journal of American Folk-Lore* 36, no. 142 (October–December 1923): 305.

175 **More than two hundred miles per hour** Tom Harpole, "Falling with the Falcon," *Smithsonian Magazine*, March 2005, https://www.smithsonianmag.com/air-space-magazine/falling-with-the-falcon-7491768.

176 **"The hawk watches"** Floyd Bralliar, *Knowing Birds Through Stories* (New York: Funk & Wagnalls, 1922), 139.

177 **"Wherever the route"** Pete Dunne, *Hawks in Flight* (Boston: Houghton Mifflin, 1988), 76.

178 **"The sparrow hawk resembles"** Winsor Marrett Tyler in Arthur Cleveland Bent, *Life Histories of North American Birds of Prey, Part 2* (1938; repr., New York: Dover, 1961), 117.

179 **"The grasshopper is held"** John W. Sugden in Bent, *Life Histories of North American Birds of Prey, Part 2*, 123.

CHAPTER 16: THE SKY ELF

190 **"Overpowering importance of neighborhood"** "Life would be twice or ten times life, if spent with wise and fruitful companions," he adds. Ralph Waldo Emerson, "Considerations by the Way," *Essays and Lectures* (New York: Penguin, 1983), 1094.

191 **"Most beautiful of all the genus"** George A. Jobanek and David B. Marshall, "John K. Townsend's 1836 Report of the Birds of the Lower Columbia River Region, Oregon and Washington," *Northwestern Naturalist* 73, no. 1 (Spring 1992): 8, https://doi.org/10.2307/3536564.

191 **"The most beautiful swallow"** John James Audubon, *The Birds of America Vol.* 1 (1840; repr., New York: Dover, 1967), 186.

191 **"If we lavished any superlatives"** Dawson, *Birds of California*, 544.

192 **Describes their face as elfin** Pete Dunne, *Essential Field Guide Companion.*

192 **"The Lay of Nimrodel"** J. R. R. Tolkien, *The Lord of the Rings: The Fellowship of the Ring*, (London: Folio Society, 1977), 385.

193 **"We are not yet"** Dawson, *Birds of California*, 545.

193 **More elevated altitudes** C. R. Brown, A. M. Knott, and E. J. Damrose, "Violet-green Swallow (*Tachycineta thalassina*)," *Birds of the World* (March 2020), https://doi.org/10.2173/bow.vigswa.01.

194 **Sibley field guides** David Allen Sibley, *Sibley Birds West* (New York: Knopf, 2016), 305.

194 **Violet-greens' wings** The relatively long length of their wings can also be validated by looking at their measurements. By the *Birds of the World* numbers, tree swallows, for instance, are about 40 percent heavier and have roughly corresponding tail lengths that are about 35 percent longer than those of violet-greens. Their wings, however, average only 4 percent longer and can overlap in length with those of their smaller cousins. C. R. Brown et al., "Violet-Green Swallow, *Birds of the World.*

194 **As Shelley did** "To a Sky-lark," in Percy Bysshe Shelley, *Poems and Prose* (London: J. M. Dent, 1995), 216–19.

ABOUT THE AUTHOR

Jack Gedney is also the author of *The Private Lives of Public Birds: Learning to Listen to the Birds Where We Live* and a compact field guide to the trees of the San Francisco Bay Area. Since 2018, he has written a column on local birds for the *Marin Independent Journal*. Jack currently co-owns a wild bird feeding and nature shop in Novato, California.

A NOTE ON TYPE

This book is set in Freight Text Pro. Created by Joshua Darden, Freight Text Pro is part of the Freight Collection, a range of integrated typeface families inspired by eighteenth-century Dutch type. The chapter titles, section titles, and headings are set in Agenda, a typeface designed by Greg Thompson.